SYMPOSIUM: SWEETENERS

Food Science and Technology

BAKERY TECHNOLOGY AND ENGINEERING, 2ND EDITION *Matz*
BREAD SCIENCE AND TECHNOLOGY *Pomeranz and Shellenberger*
CHOCOLATE, COCOA AND CONFECTIONERY *Minifie*
COOKIE AND CRACKER TECHNOLOGY *Matz*
ECONOMICS OF FOOD PROCESSING *Greig*
ECONOMICS OF NEW FOOD PRODUCT DEVELOPMENT *Desrosier and Desrosier*
FOOD ANALYSIS: THEORY AND PRACTICE *Pomeranz and Meloan*
FOOD ANALYSIS LABORATORY EXPERIMENTS *Meloan and Pomeranz*
FOOD OILS AND THEIR USES *Weiss*
FOOD PACKAGING *Sacharow and Griffin*
FOOD SCIENCE, 2ND EDITION *Potter*
FREEZING PRESERVATION OF FOODS, 4TH EDITION, VOLS. 1, 2, 3, AND
    4 *Tressler, Van Arsdel and Copley*
HANDBOOK OF SUGARS *Junk and Pancoast*
LABORATORY MANUAL FOR FOOD CANNERS AND PROCESSORS, 3RD
    EDITION, VOLS. 1 AND 2 *National Canners Association*
PHOSPHATES IN FOOD PROCESSING *DeMan and Melnychyn*
PRACTICAL BAKING, 2ND EDITION *Sultan*
PRACTICAL BAKING WORKBOOK *Sultan*
PRACTICAL FOOD MICROBIOLOGY AND TECHNOLOGY, 2ND EDITION *Weiser,
    Mountney and Gould*
PRINCIPLES OF PACKAGE DEVELOPMENT *Griffin and Sacharow*
PROCEEDINGS OF THE XIII INTERNATIONAL CONGRESS OF REFRIGERATION
    VOLS. 1, 2, 3, AND 4 *Pentzer*
QUALITY CONTROL FOR THE FOOD INDUSTRY, 3RD EDITION, VOLS. 1 AND
    2 *Kramer and Twigg*
SYMPOSIUM: SEED PROTEINS *Inglett*
THE TECHNOLOGY OF FOOD PRESERVATION, 3RD EDITION *Desrosier*

Food Service

CONVENIENCE AND FAST FOOD HANDBOOK *Thorner*
FOOD SANITATION *Guthrie*
MEAT HANDBOOK, 3RD EDITION *Levie*

Nutrition and Biochemistry

CARBOHYDRATES AND THEIR ROLES *Schultz, Cain and Wrolstad*
CHEMISTRY AND PHYSIOLOGY OF FLAVORS *Schultz, Day and Libbey*
PROTEINS AND THEIR REACTIONS *Schultz and Anglemier*
PROTEINS AS HUMAN FOOD *Lawrie*

SYMPOSIUM: SWEETENERS

Edited by George E. Inglett, Ph.D.
Chief, Cereal Properties
Laboratory, Northern Regional
Research Laboratory, ARS, USDA,
Peoria, Illinois

WESTPORT, CONNECTICUT
THE AVI PUBLISHING COMPANY, INC.
1974

Library of Congress Catalog Card Number: 73-94092
ISBN-0-87055-153-1

Printed in the United States of America

Contributors

DR. CHARLES I. BECK, Manager, Food Research, Searle Biochemics Division, G. D. Searle and Company, 2634 South Clearbrook Drive, Arlington Heights, Illinois.

DR. KARL M. BECK, Manager, Product Planning and Technical Service, Abbott Laboratories, Chemical Marketing Division, North Chicago, Illinois.

DR. LLOYD M. BEIDLER, Professor, Department of Biological Science, The Florida State University, Tallahassee, Florida.

MS. G. A. BROOKS, Chemist, CPC International, Inc., Moffett Technical Center, P.O. Box 345, Argo, Illinois.

DR. P. K. CHANG, Director of Research, International Sugar Research Foundation, Inc., 7316 Wisconsin Avenue, Bethesda, Maryland.

DR. JOHN D. COMMERFORD, Director of Technical Development, Corn Refiners Association, Inc., 1001 Connecticut Avenue, N.W., Washington, D.C.

DR. MARVIN K. COOK, Consulting Chemist, 1992 Windsor Street, Westbury, New York.

DR. FRANK R. DASTOLI, Director of Research, Miralin Corporation, Eight Kane Industrial Drive, Hudson, Massachusetts.

MR. PHILIP H. DERSE, President, WARF Institute, Inc., P.O. Box 2599, 506 North Walnut Street, Madison, Wisconsin.

MR. BRUNO GENTILI, Research Chemist, Fruit and Vegetable Chemistry Laboratory, U.S. Department of Agriculture, Pasadena, California.

MR. B. HARRY GOMINGER, Vice President and General Manager, MacAndrews and Forbes Company, Third Street and Jefferson Avenue, Camden, New Jersey.

DR. DONALD L. HARRIS, Pathologist, WARF Institute, Inc., P.O. Box 2599, 506 North Walnut Street, Madison, Wisconsin.

MR. ROBERT J. HARVEY, President, Miralin Corporation, Eight Kane Industrial Drive, Hudson, Massachusetts.

MR. JOHN E. HODGE, Research Leader, Northern Regional Research Laboratory, U.S. Department of Agriculture, Peoria, Illinois.

DR. ROBERT M. HOROWITZ, Research Chemist, Fruit and Vegetable Chemistry Laboratory, U.S. Department of Agriculture, Pasadena, California.

DR. GEORGE E. INGLETT, Chief, Cereal Properties Laboratory, Northern Regional Research Laboratory, U.S. Department of Agriculture, Peoria, Illinois.

DR. JOHN E. LONG, Assistant Director, Developmental Research, CPC International, Inc., Moffett Technical Center, P.O. Box 345, Argo, Illinois.

DR. ROBERT H. MAZUR, Research Fellow, Chemical Research Department, Searle Laboratories Division, G. D. Searle and Co., Skokie, Illinois.

DR. PAUL O. NEES, Head, Biological Department, WARF Institute, Inc., P.O. Box 2599, 506 North Walnut Street, Madison, Wisconsin.

DR. JOHN M. NEWTON, Technical Assistant to Vice President /Sales, Clinton Corn Processing Company, Clinton, Iowa.

PROFESSOR ROSE MARIE PANGBORN, Department of Food Science and Technology, University of California, Davis, California.

MR. MERRILL O. TISDEL, Chief, Chronic Toxicology, WARF Institute, Inc., P.O. Box 2599, 506 North Walnut Street, Madison, Wisconsin.

DR. J. CLYDE UNDERWOOD, Research Leader, Eastern Regional Research Laboratory, 600 East Mermaid Lane, Philadelphia, Pennsylvania.

MR. H. VAN DER WEL, Biochemist, Unilever Research Laboratorium, Olivier van Noortlaan 120, Vlaardingen, The Netherlands.

PROFESSOR A. J. VLITOS, Chief Executive, Group Research and Development, Tate and Lyle, Limited, Philip Lyle Memorial Research Laboratory, P.O. Box 68, Reading, Berks., RG6 2BX, England.

DR. BRUCE J. WALTER, Agricultural Economist, Horticultural and Special Crops Branch, Marketing Economics Division, ERS, USDA, Washington, D.C.

MR. E. K. WARDRIP, Manager, Product Development and Research, Clinton Corn Processing Company, Clinton, Iowa.

DR. M. O. WARNECKE, Section Leader, CPC International, Inc., Moffett Technical Center, P.O. Box 345, Argo, Illinois.

DR. JONATHAN W. WHITE, JR., Chief, Plant Products Laboratory, Eastern Regional Research Laboratory, 600 East Mermaid Lane, Philadelphia, Pennsylvania.

DR. FERDINAND B. ZIENTY, Manager, Research and Development, Food and Fine Chemicals, Monsanto Company, 800 North Lindbergh Boulevard, St. Louis, Missouri.

Preface

This book contains selected papers from a Symposium on Sweeteners held at the American Chemical Society meeting in Dallas, Texas, on April 9-13, 1973. Reviewed at this Symposium, sponsored by the Agricultural and Food Chemistry and Carbohydrate Divisions, were the important aspects of natural and synthetic sweeteners. Areas of current interest and the latest technological progress received particular emphasis.

The multidisciplinary study of sweeteners delineated in this book includes the fields of biochemistry, chemistry, economics, engineering, food science, molecular biology, nutrition, physiology, technology, and toxicology. Contributors to this volume are authorities having a close and continuing acquaintance with the science and technology of sweeteners.

Both subjects and authors were selected by the Editor to cover the broad areas concerned with natural and synthetic sweeteners. Another editor and Symposium organizer could have easily selected differently. My intention was to provide information on sweeteners that would be useful to all people concerned with any aspect of their manufacture and application.

The Editor expresses his utmost appreciation to those Symposium participants who were willing to meet his challenge and to their companies, government agencies, and universities who allowed their participation.

The Editor is also indebted to the officers of the Agricultural and Food Chemistry Division of the American Chemical Society; namely, Drs. E. L. Wick, S. Kazeniac, and R. J. Magee, who allowed us the forum for this Symposium.

Associates who gave substantial help throughout the editing of this book are: Wilma J. Bailey, Charles W. Blessin, Linda L. Borror, Doris M. Davis, Robert J. Dimler, John E. Hodge, Wilbur C. Schaefer, and Virginia Mae Thomas. Their services are duly acknowledged.

Career-service employees of the U.S. Department of Agriculture are expressing their own opinions in their contributed articles, and not those of the United States Government.

GEORGE E. INGLETT

May 21, 1973

Contents

CHAPTER 1

G. E. Inglett | Sweeteners: New Challenges and Concepts

The major sweeteners, worldwide, are sucrose and corn sugars. Sucrose is one of the oldest sweeteners known to man, and references to it go back as far as the earliest written records. Without doubt, it is the oldest sweetener derived from commercially cultivated plants. Sucrose is widely distributed in a great variety of plants, but only four — sugarcane, sugar beets, palm trees, and maple trees — are commercial sources of the granular product. Nearly all the world's commercial supply of sucrose, however, comes from sugarcane and sugar beets. Recovering sucrose from sugar beets is continuous from the beet to the finished product. Cane sugar is generally produced as a raw sugar in cane-growing areas and refined in the country where it is to be marketed.

Dextrose, except that which forms half of invert sugar, and corn syrup, or dextrose syrup, are derived from starch by hydrolysis. Starch suitable for the manufacture of dextrose and syrup may come from any one of numerous plant sources, although corn is the most important source in the U.S. The corn wet-milling industry practices a considerable degree of flexibility in making changes in response to the varying market requirements for starch, dextrose, and corn syrup. This industry converts almost 60% of domestic corn starch into dextrose and corn syrups (Senti 1965).

The starch-processing industry has undergone a dynamic change in the past decade with the advent of enzymatic dextrose production, using the enzyme glucoamylase. Glucoamylase is produced by several *Aspergillus* species, together with an undesirable enzyme, trans-glucosidase. Transglucosidase is removed by clay mineral (Kooi *et al.* 1962; Inglett 1963) or other procedures (Kerr 1961; Hurst and Turner 1962A, 1962B; Kathrein 1963). Glucoamylase is produced by submerged-culture fermentation in the general manner developed by USDA's Northern Regional Research Laboratory (LeMense and Van Lanen 1948; Smiley *et al.* 1964). Ultraviolet-induced mutation of *Aspergillus* species has greatly increased glucoamylase production with substantially less transglucosidase per glucoamylase unit. An extensive discussion of dextrose processing was recorded by Kooi and Armbruster (1967).

Enzymatic starch hydrolysis gives higher yields and greatly reduces evaporation costs, because a 30 to 40% starch concentration can be hydrolyzed compared with 12 to 20% by the acid-conversion process. The hydrolyzate is further processed with activated charcoal and ion-exchange resins to produce food-grade syrups and dextrose.

A challenging area of sweetener development is the preparation of various new kinds of corn syrup (Barfoed 1967; Kossoy 1968). The functional properties of syrups are particularly important. However, the latest developments involve sweet-taste qualities. Most commercial corn syrups do not have the sweetening power of sucrose on an equal dry weight basis. Larger markets for corn syrup are considered possible if it could be made sweeter (Ballinger and Larkin 1964).

Levulose (fructose) is at most 1.7 times as sweet as sucrose. Enzymatic methods for isomerizing dextrose to fructose (levulose) have been actively pursued since the studies of Marshall and Kooi (1957) and Marshall (1960). Many Japanese workers have published their results on dextrose-isomerizing enzyme systems (Tsumura *et al.* 1967; Danno 1971; Tisuka *et al.* 1971; Takasaki and Tanabe 1971). Brownewell (1971) and Lee *et al.* (1972) also developed processes for preparing glucose isomerases. A method of making isomerized glucose syrups has been patented by the Clinton Corn Processing Company (Cotter *et al.* 1971). The properties of their commercial

TABLE 1.1

PER CAPITA CONSUMPTION OF THE MAJOR SWEETENERS
IN THE U.S., 1960-71

	Per Capita Consumption[1]		
Year	Sucrose Beet and Cane	Corn Sugar	Corn Syrup
1960	97.6	3.7	10.1
1961	97.7	3.7	10.6
1962	97.2	3.9	11.5
1963	96.6	4.5	12.3
1964	96.5	4.4	13.6
1965	96.4	4.5	13.6
1966	97.4	4.5	14.0
1967	97.1	4.6	14.0
1968	99.8	4.7	14.7
1969	99.9	4.8	15.0
1970	102.2	5.0[2]	15.8
1971	101.6[2]	5.2[2]	16.2

[1] Source: Anon. 1972; lb per person.
[2] Preliminary data.

high-fructose corn syrups are revealed by Wardrip (1971); also see Chapter 8. Other companies in the U.S. are actively investigating the preparation of levulose syrups from dextrose (Kooi and Smith 1972). Applications are being made of immobilized glucose isomerase (Strandberg and Smiley 1971). Research activity also continues on the chemical conversion of dextrose to fructose with such catalysts as alkali-metal aluminate (Haack *et al.* 1966).

In 1965, an estimated 18.7 billion lb of assorted sugars worth nearly 2 billion dollars went into foods (Ballinger and Larkin 1964). In the U.S., sucrose is consumed in the largest quantities of all sweeteners, amounting to 101.6 lb per capita in 1971 (Anon. 1972). The per capita consumption of dextrose was 5.2 lb and of corn syrup 16.2 lb for that year. Per capita consumption of the major sweeteners in the U.S. from 1960 to 1971 is given in Table 1.1. The per capita use of corn syrup rose by 6.1 lb between 1960 and 1971, an increase of 60%. The per capita use of dextrose increased by 1.5 lb between 1960 and 1971, an increase of 40%. The increase in dextrose and corn syrup consumption illustrates the important changes that become apparent over an extended period of time.

SWEETENERS AND HEALTH

The relationship of consumption and health is beyond the scope of this volume; however, the importance of this relationship should not be underestimated. Not only should the major sweeteners continue to undergo rigorous scientific scrutiny, but the other natural and synthetic sweeteners should also be considered with suspicion until proved absolutely harmless.

The natural exotic sweeteners, such as stevioside and glycyrrhizin, have a history of use by various people, generally of a given geographic location and of low economic status. These natural products do not exhibit any sensational toxicity like strychnine, for example, but long-term toxicity effects remain to be determined. Toxicity of many natural products under conditions of food use over a long period in humans is difficult, if not impossible, to evaluate because of human habits and genetic and aging differences. This challenge is still to be met.

NEW CONCEPTS OF SWEETENERS

Sweetness is a gustatory response invoked by substances on sweet taste buds which transmits a message to the brain indicating sweet taste. The chemical nature of various substances that excite a sweet taste has been extensively studied with no universally accepted explanation. This challenge of sweeteners is one with great potential. More basic information is needed concerning the nature of sweetness in terms of both molecular biology and physiology. Although much has been said and written about sweetness, basic experimental data are still fragmentary.

Macromolecular Taste Modifier

An important approach to taste perception is the study of the strange properties of the miracle fruit (Fig. 1.1). Although this

Courtesy of G. D. Inglett, Peoria, Ill.
FIG. 1.1. MIRACLE FRUIT, *SYNSEPALUM DULCIFICUM*
(SCHUM. ET THONN.)

miracle fruit (*Synsepalum dulcificum*) has been known in the literature since 1852 (Daniell 1852) to cause sour foods to taste sweet. Scientific investigations of the fruit were not made until Inglett and his associates (1965) found that the active principle was macromolecular, with some experimental evidence that it might be a glycoprotein. Subsequently, Brouwer *et al.* (1968), Henning *et al.* (1969), and Kurihara and Beidler (1968) all reported that the active principle was a glycoprotein with a molecular weight of 42,000. This taste-modifying substance revealed a new concept of taste perception generally, and of sweet taste particularly. Until this time, only small molecules were considered as sweet-invoking substances. This was the first time that macromolecules were considered capable of participating in either taste perception or modification of taste (Inglett *et al.* 1964; Inglett 1970, 1971A, 1971B, 1971C).

Macromolecular Sweeteners

Further research on exotic sweeteners of natural origin gave added insight and new information about this new concept. Scientific recognition of the supersweetness of the fruit of *Dioscoreophyllum cumminsii*, called the serendipity berry (Inglett and May, 1968), was to spur on new findings (Fig. 1.2). The macromolecular properties of

FIG. 1.2. SERENDIPITY BERRIES, *DIOSCOREOPHYLLUM CUMMINSII* DIELS

the *D. cumminsii* sweetener was found and reported initially in 1967 (Inglett and Findley 1967; Inglett and May 1969). Research workers at Monell Senses Center and Unilever Research Laboratorium independently confirmed the protein nature of the serendipity berry sweetener (Morris and Cagan 1972; van der Wel 1972). It thus seems appropriate to call the sweetener serendip.

Besides studies on miracle fruit and the serendipity berry, a large variety of plant materials were examined systematically by Inglett and May (1968: Inglett 1971D) for sweetness intensity and quality. Another African fruit containing an intense sweetener was katemfe, or the miraculous fruit of the Sudan (Daniell 1855). Botanically the plant is *Thaumatococcus daniellii*. Inside the fruit three large black seeds are surrounded by a transparent jelly and a light yellow aril at the base of each seed. The mucilaginous material around the seeds is intensely sweet and causes other foods to taste sweet. The seeds were observed in trading canoes in West Africa as early as 1839, and were reported to be used to sweeten bread, fruits, palm wine, and tea. Preliminary studies have indicated a substance similar to the serendipity berry sweetener (Inglett and May 1968). In 1972, van der Wel reported the active principle of *T. daniellii* Benth fruit (katemfe) to be a protein (see Chapter 17).

Sweetness Probe

The macromolecules responsible for sweetness either by modification of taste (miracle fruit) or direct sweetness (serendipity berry, katemfe) without question must have a portion of their structure that is essential for the sweet taste. The author calls this center a "sweetness probe". The challenge of finding sweetness probes and of demonstrating their structures and utility is still before us.

The sweetness probe of the miracle fruit's active principle may lie in its bonding between the carbohydrate moiety and the protein chain. This chemical bonding could be considered analogous to the aglycone glycoside binding of such intense sweeteners as stevioside, osladin, and glycyrrhizin. The induced sweetness of the glycoprotein in miracle fruit is attributed to the sugar portion of the molecule, according to Beidler (1971). The protein portion holds onto the receptor site, placing the arabinose or xylose sugars close enough for inducing sweet taste. Although this assumption may seem to be reasonable, the necessity of protons for sweet taste needs additional investigation. Perhaps protons modify the conformation of the glycoprotein to give the necessary shape to induce sweet taste. If the sweetness probe can be chemically defined, analogs of low molecular weight may be synthesized that would not have the residual taste deficiency of the macromolecule.

The serendipity berry sweetener, which is a protein (Morris and Cagan 1972; van der Wel 1972), may have a different type of sweetness probe. Since the *D. cumminsii* sweetener acts directly on the taste buds as a probe, a peptide linkage analogous to the aspartic acid sweeteners (Mazur *et al.* 1969), to be discussed in this volume, may be an essential factor. At this stage of development, an aspartic acid peptide probe center cannot be excluded. More information is needed on the active principles of miracle fruit, serendipity berry, and katemfe.

As an extension of the sweetness probe theory, the chemical structures of the intense sweeteners — stevioside, glycyrrhizin, osladin, and the dihydrochalcones — are examined. The obvious similarity among these supersweeteners is the occurrence of a $(1'\rightarrow2)$-oxygen-linked disaccharide in the glycoside attached to an aglycone. At the Northern Regional Research Laboratory, a program was initiated to synthesize $(1'\rightarrow2)$-oxygen-linked carbohydrates; preliminary information on this work will be found in Chapter 20.

Enzymatically Modified Sweeteners

Although a large number of substances exhibit some resemblance to sucrose-type sweetness (Inglett 1971C), only a few can be considered suitable for human consumption. The building blocks of foodstuffs — carbohydrates, proteins, lipids — are among the substances generally recognized as safe. Sweeteners derived from these materials may be less hazardous in foods than totally foreign organic structures. The protein sweeteners of serendipity berry and katemfe should be considered with this prospect in mind.

The protein sweeteners also lend themselves to enzymatic modifications. A successful modification could give a product with improved sensory and enhanced stability properties. Proteolytic enzyme treatment of the serendipity berry sweetener (Inglett and Findley 1967; Inglett and May 1969) appeared to yield a low molecular weight sweetener. By successive enzymatic degradation of *Dioscoreophyllum cumminsii* sweetener with pectinase and bromelain and (or) papain, a polypeptide (mol wt 6000) was obtained having a sweetness of 800 to 3200 times that of sucrose (Parker 1973). The structures and properties of this sweetener, as well as of the unmodified ones, need additional study.

SUMMARY

Challenges facing the sweetener industry are: (1) health and safety of natural and synthetic sweeteners; (2) sweeter syrups from corn; (3) the need for a superior synthetic sweetener; and (4) a better scientific understanding of sweetness.

The concept that a sweet probe portion of some macromolecules is responsible for invoking a sweet taste is expanded. The sweet probe concept is discussed and illustrated by the active taste principles of miracle fruit (*Synsepalum dulcificum*), serendipity berry (*Dioscoreophyllum cumminsii*), katemfe (*Thaumatococcus daniellii*), and $(1'\rightarrow2)$-oxygen linked disaccharide sweeteners (stevioside, osladin, glycyrrhizin, and the dihydrochalcones).

The challenge of finding an innocuous intense sweetener with the taste quality of sucrose and with the pharmacology of water is a formidable objective. The pitfalls are numerous and the hazards, great; but success will be sweet!

8 SWEETENERS

BIBLIOGRAPHY

ANON. 1972. Agricultural statistics. U.S. Dep. Agr., Washington, D.C.
BALLINGER, R. A., and LARKIN, L. C. 1964. Sweeteners used by food processing industries. U.S. Dep. Agr., Econ. Res. Serv., Agr. Econ. Rep. No. 48, Washington, D.C.
BARFOED, VAN H. 1967. Enzymatic dextrose and corn syrup. Staerke 19, 2-8. (German).
BEIDLER, L. M. 1971. The physical basis of the sweet response. In Sweetness and Sweeteners. G. G. Birch. L. F. Green, and C. B. Coulson (Editors). Applied Science Publishers, Ltd., London.
BROUWER, J. N., VAN DER WEL, H., FRANCKE, A., and HENNING, G. J. 1968. Miraculin, the sweetness-inducing protein from miracle fruit. Nature 220, 373-374.
BROWNEWELL, C. E. 1971. Process for production of glucose isomerase. U.S. Pat. 3,625,828. December 7.
COTTER, W. P., LLOYD, N. E., and HINNAU, C. W. 1971. Method for isomerizing glucose syrups. U.S. Pat. 3,623,953. November 30.
DANIELL, W. F. 1852. Miraculous berry of Western Africa. Pharm. J. 11, 445-448.
DANIELL, W. F. 1855. Katemfe, or the miraculous fruit of Sudan. Pharm. J. 14, 158-159.
DANNO, G. 1971. Studies on D-glucose-isomerizing enzyme from Bacillus coagulans strain HN-68. Part VI. The role of metal ions in the isomerization of D-glucose and D-xylose by the enzyme. Agr. Biol. Chem. 35, 997-1006.
HAACK, E., BRAUN, F., and KOHLER, K. 1966. Process for the manufacture of D-fructose. U.S. Pat. 3,256,270. June 14.
HENNING, G. J., BROUWER, J. N., VAN DER WEL, H., FRANCKE, A. 1969. Miraculin, the sweetness-inducing principle from miracle fruit. In Olfaction and Taste. III. C. Pfaffman (Editor). Rockefeller University Press, New York.
HURST, T. L., and TURNER, A. W. 1962A. Method of refining amyloglucosidase. U.S. Pat. 3,047,471. July 31.
HURST, T. L., and TURNER, A. W. 1962B. Method of refining amyloglucosidase. U.S. Pat. 3,067,108. December 4.
INGLETT, G. E. 1963. Purification and recovery of fungal amylases. U.S. Pat. 3,101,302. August 20.
INGLETT, G. E. 1970. Natural and synthetic sweeteners. Hort. Science 5, 139-141.
INGLETT, G. E. 1971A. Intense sweetness of natural origin. In Sweetness and Sweeteners. G. G. Birch, L. F. Green, and C. B. Coulson (Editors). Applied Science Publishers, Ltd., London.
INGLETT, G. E. 1971 B. Miracle Fruit. Botanicals. P.O. Box 3034, Peoria, Ill.
INGLETT, G. E. 1971C. Recent Sweetener Research. 2nd ed. Botanicals. P.O. Box 3034, Peoria, Ill.
INGLETT, G. E. 1971D. Serendipity Berries. Botanicals. P.O. Box 3034, Peoria, Ill.
INGLETT, G. E., DOWLING, B., ALBRECHT, J. J., and HOGLAN, F. A. 1964. A new concept in sweetness-taste modifying properties of miracle fruit. (Synsepalum dulcificium). 148th Am. Chem. Soc. Meeting, Div. Agr. Food Chem., Paper 1. Chicago, Ill.
INGLETT, G. E., DOWLING, B., ALBRECHT, J. J., and HOGLAN, F. A. 1965. Taste-modifying properties of miracle fruit (Synsepalum dulcificum). J. Agr. Food Chem. 13, 284-287. .
INGLETT, G. E., and FINDLAY, J. C. 1967. Serendipity berry — source of a new macromolecular sweetener. 154th Am. Chem. Soc. Meeting, Div. Agr. Food Chem., Paper 75, Chicago, Ill.
INGLETT, G. E., and MAY, J. F. 1968. Tropical plants with unusual taste properties. Econ. Bot. 22, 326-331.

INGLETT, G. E., and MAY, J. F. 1969. Serendipity berries (*Dioscoreophyllum cumminsii*) — source of a new intense sweetener, J. Food Res. *34*, 408-411.

KATHREIN, H. R. 1963. Treatment and use of enzymes for the hydrolysis of starch. U.S. Pat. 3,108,928. October 29.

KERR, R. W., 1961. Method of making dextrose using purified amyloglucosidase. U.S. Pat. 2,970,086. January 31.

KOOI, E. R., and ARMBRUSTER, F. C. 1967. *In:* Starch: Chemistry and technology. Vol. II: Industrial aspects. Roy L. Whistler and Eugene F. Paschall (Editors). Academic Press, New York.

KOOI, E. R., HARJES, C. F., and GILKISON, J. S. 1962. Treatment and use of enzymes for the hydrolysis of starch. U.S. Pat. 3,042,584. July 3.

KOOI, E. R., and SMITH, R. J. 1972. Dextrose-levulose syrup from dextrose. Food Technol. *26*, No. 9, 57-60.

KOSSOY, M. W. 1968. The function of corn syrups in jams, jellies, and preserves. Food Prod. Develop. *1*, 35.

KURIHARA, K., and BEIDLER, L. M. 1968. Taste-modifying protein from miracle fruit. Science *161*, 1241-1243.

LEE, C. K., HAYES, L. E., and LONG, M. E. 1972. Process of preparing glucose isomerase. U.S. Pat. 3,645,848. February 29.

LEMENSE, E. H., and VAN LANEN, J. M. 1948. Process for preparing starch hydrolyzing enzyme with *Aspergillus*. U.S. Pat. 2,451,567. Oct. 19.

MARSHALL, R. O. and KOOI, E. R. 1957. Enzymic conversion of D-glucose to D-fructose. Science *125*, 648-649.

MARSHALL, R. O. 1960. Enzymatic process. U.S. Pat. 2,950,228. August 23.

MAZUR, R. H., SCHLATTER, J. M., and GOLDKAMP, A. H. 1969. Structure-taste relationships of some dipeptides. J. Am. Chem. Soc. *91*, 2684-2691.

MORRIS, J. A., and CAGAN, R. H. 1972. Purification of monellin, the sweet principle of *Dioscoreophyllum cumminsii*. Biochim. Biophys. Acta *261*, 114-122.

PARKER, K. J. 1973. Sweetening agent for foods from *Dioscoreophyllum cumminsii*. Ger. Offen 2,224,644, Feb. 1; CA 78, 109547a.

SENTI, F. R. 1965. The industrial utilization of cereal grains. Cereal Sci. Today *10*, 320-327; 361-362.

SMILEY, K. L., CADMUS, M. C., HENSLEY, D. E., and LAGODA, A. A. 1964. High-potency amyloglucosidase-producing mold of the *Aspergillus niger* group. Appl. Microbiol. *12*, 455.

STRANDBERG, G. W., and SMILEY, K. L. 1971. Free and immobilized glucose isomerase from *Streptomyces phaeochromogenes*. Appl. Microbiol. *21*, 588-593.

TAKASAKI, Y., and TANABE, O. 1971. Enzymatic method for converting glucose syrups to fructose. U.S. Pat. 3,616,221. October 26.

TISUKA, H., AYUKAWA, Y., SUEKANE, M., and KANNO, M. 1971. Production of extracellular glucose isomerase by *Streptomyces*. U.S. Pat. 3,622,463. November 23.

TSUMURA, N., HAGI, M., and SATO, T. 1967. Enzymatic conversion of D-glucose to D-fructose. Pt. IX. Dehydration and preservation of the cell as enzyme source. Agr. Biol. Chem. *31*, 908.

VAN DER WEL, H. 1972. Isolation and characterization of the sweet principle from *Dioscoreophyllum cumminsii* (Stapf) Diels. FEBS Lett. *21*, No. 1, 88-90.

WARDRIP, E. K. 1971. High-fructose corn syrup. Food Technol. *25*, 501-503.

CHAPTER 2

Lloyd M. Beidler | Biophysics of Sweetness

The importance of this volume centers on the fact that sweetness is a quality of food that is greatly preferred by most humans. The origin of this preference is not known, although it may appear as early as the fourth fetal month (Bradley and Mistretta 1972). Swallowing by the human fetus increases when saccharin is injected into the amniotic fluid and decreases in response to a bitter substance (De Snoo 1937; Liley 1971). Sweet preference is also evident in babies several days after birth (Desor *et al.* 1972). This preference is often maintained into adult life and has a number of deleterious effects, particularly dental caries and obesity.

How can we satisfy our craving for sweets without suffering adverse consequences? The answer is to develop nontoxic molecules that serve as sweet stimuli but do not appreciably affect the body's metabolism. The taste sensation should be a pleasant sweetness without side tastes, and should appear and disappear quickly as the stimulus is presented and withdrawn. Sucrose meets most of these requirements, but its caloric content is unacceptable. Substitutes, such as saccharin, dulcin, and cyclamate, do not have the pleasant sweet quality of sucrose and their toxicity may be questioned. Aspartylphenylalanine methyl ester has an excellent quality of sweetness, but there remains the problem of amino acid imbalance. Miraculin has a good quality of sweetness, is not toxic, but can only be used for special purposes. Monellin quality is moderate but produces a lingering sweetness. Today there is a hurried search for more acceptable molecules.

BIOPHYSICS OF TASTE RECEPTORS

Taste Receptor Morphology

The basis of sweetness is the ability of a molecule to properly interact with certain taste receptors. It is the purpose of this chapter to discuss our present understanding of this interaction.

Man possesses up to half a million taste cells, clustered into groups of 40 to 60 to form taste buds. The apical end of the taste receptors consists of microvilli (0.2μ x 2.0μ, fingerlike projections) which are in direct contact with the saliva (Fig. 2.1). The remainder of the cell's surface is protected by flat nonpermeable cells that cover the

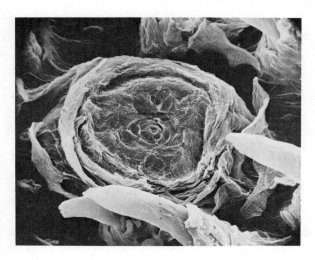

FIG. 2.1 TOP VIEW OF RAT FUNGIFORM PAPILLA
WITH A TASTE BUD PENETRATING THE SURFACE
NEAR THE CENTER

surface of the tongue (Mistretta and Beidler 1967; Mistretta 1971). Thus, the interaction of taste stimulus with receptor most probably occurs at the microvilli, although intravenous stimulation can occur with certain substances (Bradley 1972).

The taste cells live only about a week and a half and are continually replaced (Beidler and Smallman 1965). It is quite possible that the microvilli may be replaced even faster. Since it is known that individual protein molecules within cell membranes are constantly replaced, it can be assumed that the same is true of taste cells. Thus, the cell with which the sweet stimulus interacts is very dynamic and ever-changing.

Receptor Membrane

Cell membranes are constructed of a bimolecular layer of phospholipids radially arranged, with proteins covering the inner and outer surfaces. The total thickness is about 100 Å. The plasma membrane is not entirely uniform, however, but has a mosaic character: glycoproteins dip into the phospholipid layer with their hydrophobic portions and their hydrophilic components are arranged outwardly. In addition, some cell membranes, such as the red cell, are loosely covered with polysaccharides (Fig. 2.2). There is no reason to expect that the general structure of the taste cell membrane differs greatly from that of other membranes.

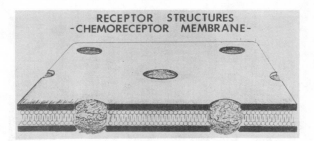

FIG. 2.2. SCHEMATIC OF A MODEL OF THE TASTE
RECEPTOR MEMBRANE CONSISTING OF A BI-
MOLECULAR PHOSPHOLIPID LAYER COVERED ON
BOTH SIDES BY PROTEIN
Membrane surface interrupted by protein-filled pores.

Tens of thousands of different molecules can bind reversibly to
the taste receptor membrane. Alkali cations play a dominant role in
salty taste (Beidler 1954, 1971) and hydrogen ions in sourness.
However, not all bound hydrogen ions lead to sourness; in fact, most
such ions are inaffective. This is an indication that stimulus binding
alone (affinity) is not affective, but must be followed by a local
change in membrane properties, probably a conformational change
(intrinsic activity). Binding of hydrogen ions to the proper
membrane protein carboxyl group leads to cell excitation and
sourness. However, not all receptor sites bind cations in the same
way: some bind sodium much better than potassium, and others the
reverse. This is determined by the electric field strength or nature of
the receptor site.

Sweet and bitter molecules are much more complex than salty and
sour. They therefore display much higher specificities in their
interactions with the taste cell membranes. Furthermore, not all
sweet binding sites are identical: some react better with fructose than
glucose, but others do the opposite. Thus we may conclude that
there are many varied receptor sites on the taste microvilli
membranes that can bind taste molecules and that some of these
bindings lead to conformational changes in the membrane and
eventual taste responses.

It is possible to study the molecular structures of sweet substances
and look for common properties. Shallenberger and Acree (1967,
1971) have been must successful; they stress the fact that all sweet
molecules have an electronegative atom B and a polarized system
A-H with a distance between them of about 3 Å. Conformational
barriers can account for differences in L and D amino acids. Kier
(1972) subsequently examined amino acids in more detail, with
molecular orbital methods, and concluded that they all have the

A-H—B feature but in addition a third electron-rich site, X, that may be involved in a dispersion bonding to the taste receptor site. Their atomic pattern with approximate distances is:

These structure-taste relationships, however, are not complete enough to allow prediction of taste of new molecular structures. The lack of a homogeneous population of taste receptor sites makes this problem most difficult.

Membrane proteins are the most likely units to contain the receptor sites to which sugars bind. With this in mind several laboratories have tried to isolate biochemically protein fractions from tongue membranes to which sugars attach (Dastoli and Price 1966; Dastoli 1972). To date, these proteins have not been purified or completely characterized.

Taste cells, like all other sensory cells, are electrically charged with a difference of potential of about 0.030 volt across the membranes. Since the membrane is about 100 Å thick, this voltage gradient is about 30,000 volts/cm! When the cell is stimulated, the membrane near the nerve junction depolarizes and the voltage gradient decreases. It is this potential change that leads to excitation of the associated nerve and the creation of nerve action potentials which spread along the nerve toward the brain. The number of action potentials per unit time conveys information about the intensity of the sweetness to the brain.

Quantitative Relations

Of major concern today is how the sweet molecules react with the taste cell membrane to cause its depolarization. There are two requirements. First, the stimulus molecule must bind to a specific sweet site on the microvillus membrane. Second, this binding must bring about a proper conformational change within the membrane. The initial relationship can be quantitatively described in its simplest form as:

$$S + P \rightleftharpoons SP$$

The mass action law can be applied to reveal:

$$R = \frac{CKR_m}{1 + CK} \qquad \text{(Equation 1)}$$

where

R = magnitude of response to conc. C
R_m = magnitude of response to maximum conc.
K = association constant
C = stimulus conc.

The above relationship refers to cases where a homogeneous population of receptor sites exists and where no other interactions occur. If a bimodal population is present, then a simple alteration must be made. Another interesting case is where the number of hydrogen ions that binds to the membrane is so great that the increased positive charge decreases the affinity for another hydrogen ion to bind to a receptor site. This may be referred to as a cooperative membrane effect. Simultaneous binding of the associated anion relieves this positive charging, and thus the sourness depends not only on pH but also on the particular type of acid. Ionic strength is also important. For example, acetic acid may be buffered and its pH greatly increased without a very great alteration in sourness (Beidler 1967). The above equation can be rewritten and modified to:

$$\log \frac{R}{R_m - R} = \text{-pH -pK}_i + A$$

where A is related to the electrical charge of the membrane in the environment of the hydrogen-ion receptor site (Beidler 1967).

The magnitude of response can be objectively evaluated by recording the frequency of electrical action potentials traveling along the taste nerve of animals or man. The latter is performed during ear surgery when the chorda tympani taste nerve in the middle ear is available for recording (Fig. 2.3). Data are in agreement with the above concept of taste stimulus adsorption (Beidler 1954, 1971).

Threshold

Sweet threshold, C_T, can be expressed as:

$$C_T = \frac{1}{K R_m} \qquad \text{(Equation 2)}$$

It should be noted that K is related to the strength of affinity with which the taste stimulus is bound to its complementary sweet

RESPONSE OF HAMSTER TO SUCROSE

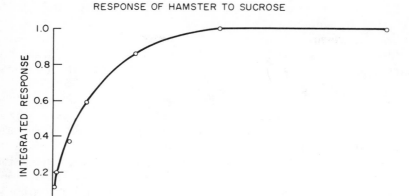

FIG. 2.3. RESPONSE OF HAMSTER TASTE RECEPTORS TO VARIOUS SUCROSE CONCENTRATIONS
Integrated response measured from neural activity of chorda tympani taste nerve.

receptor site. The logarithm of K is directly related to the binding energy, which is only several kilocalories per mole for sucrose but over three times greater for saccharin. A stimulus molecule with high affinity remains bound longer, its sweet threshold is usually lower, and taste persistency is greater. The maximum magnitude of response, R_m, is related to the intrinsic activity, or the ease with which a proper conformational change occurs when the stimulus molecule binds to the taste cell receptor site, and the number of receptor sites with which the taste molecule can interact.

It should be emphasized that the interactions of taste molecules with receptor sites are basically similar to interactions between drugs and tissues. For example, the biological and receptor-binding properties of insulin can be separated (Cuatrecasas and Illiano 1971). It is possible to change the rate of glucose transport in adipose tissue cells (intrinsic activity) without changing the number of insulin molecules bound to the cell membrane or the strength of binding (affinity). Likewise, some specific molecules that affect pancreatic cells, such as alloxan, also inhibit the sweet response of glucose (Zawalich 1972).

Much of the literature refers to intensity of sweetness in terms of threshold. This is very unfortunate, since sweetness is really an intensity of sensation that can by measured only by psychophysical techniques such as magnitude estimation (Moskowitz 1970, 1971).

In fact, the maximum sweetness of saccharin is not much greater than that of sucrose, even though its threshold is much lower. This can be appreciated by reference to Eq. 1 and 2. If saccharin and sucrose were to elicit the same intensity of sweetness (R_m) at high concentrations, and if the K for saccharin were 1000 and for sucrose 10, then according to Eq. 2, the threshold of saccharin would be 100 times less than that of sucrose. The response-concentration curve of saccharin would start at lower concentrations and rise more rapidly than that of sucrose, but the two would approach about the same response magnitude at high concentrations.

The above consideration is of some practical importance, since most theories relating chemical structure to sweetness consider only the ability to bind (affinity) and not the total transduction ability (intrinsic activity). Furthermore, threshold is often falsely used as a measure of the strength of binding.

SEARCH FOR NEW SWEETENERS

General Properties

The properties of taste receptors discussed above reveal a few principles useful in the search for new sweeteners.

(1.) The affinity of the sweet molecule should be moderately high so that only small amounts of sweetener are necessary, but not too high so that sweetness doesn't linger.

(2.) The intrinsic activity of the molecules should be as high as possible. It is not evident that any known sweetener has an intrinsic activity more than double or triple that of sucrose.

(3.) The sweetener should neither be toxic nor alter the normal metabolic state. Since any useful sweetener will be used in a very large variety of foods, this requirement is difficult to fulfill. If a molecule interacts with taste cells, there is a reasonable chance that it may also interact with other cells whose membranes may be similar. Therefore, it would be ideal to find a sweetener that can be metabolized and act as a food, but whose affinity and intrinsic activity are such that small concentrations are needed for sweetness.

The discovery in our laboratory and the subsequent confirmation by Unilever that the active ingredient of miracle fruit is a protein opened up new possibilities (Kurihara and Beidler 1968). Previously it was thought that macromolecules could not evoke a taste response. It has been recently discovered that the active ingredient of the serendipity berry is also a protein (Morris and Cagan 1972). Mazur, Schlatter and Goldkamp (1969) discovered that certain dipeptides have a pleasant sweet quality at reasonably low concentrations.

Miracle Fruit

Suitable proteins and polypeptides are a good bet for a new sweetener. Let us first consider miracle fruit. The taste molecule is a glycoprotein of about 44,000 molecular weight. It has been postulated that the sugar component interacts with the normal sweet receptor site and that the remainder of the protein molecule binds to the cell surface in such a way that the sugar component is maintained near the receptor site. A decrease in pH allows the sugar to fit into the site, and sweetness can be turned on and off by merely changing the pH. The effectiveness of the sweetening varies with the type of acid and can be further controlled by adding salts or buffers (Beidler 1967; Kurihara *et al.* 1969).

This protein has the greatest effective affinity that is known for any sweetener (Fig. 2.4). It operates at 10^{-7} M concentration and is tightly bound (Kurihara, Kruihara and Beidler 1969). This means that the molecule is bound for many minutes and that sweetness can

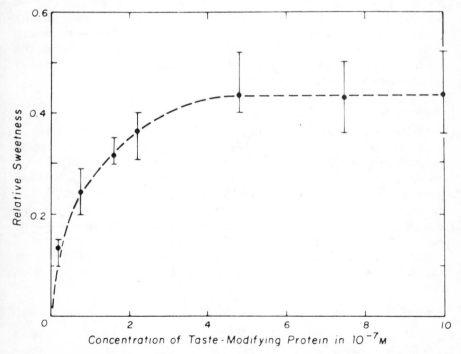

FIG. 2.4. RELATIVE SWEETNESS, MEASURED BY EQUIVALENT SWEETNESS OF SUCROSE CONCENTRATIONS (*M*), OF 0.01*M* CITRIC ACID AFTER VARIOUS CONCENTRATIONS OF MIRACLE FRUIT GLYCOPROTEIN HAVE BEEN APPLIED TO THE TONGUE

be turned on and off without adding more sweetener molecules. No other known molecule has this property. Its intrinsic activity is slightly less than that of sucrose.

The very high affinity results in very low sweetener consumption during a meal. The caloric content of this ingredient is less than 1 Calorie per meal! Its disadvantages are that application to the tongue must precede food intake, and that the meal preparation must be altered. At the present time this sweetener should be useful to diabetics and diet watchers, and can be incorporated into certain foods and confectioneries for the general public. Miralin Inc. introduced a product for market during 1973 (see Chapter 18).

Loss of activity occurs when amino groups and possibly imidazoles are modified. On the other hand, modification of the sulfhydryl and carboxylic groups and tryptophan and tyrosine residues does not result in loss of more than 25% of the normal activity (Kurihara 1972). The intrinsic activity can theoretically be increased by replacing the sugars (possibly xylose and arabinose) by fructose. The 30 sec or more latency of action should be investigated. This may be due either to its difficulty in reaching the receptor site partially buried in the membrane, or to the conformational changes necessary to provide a good fit. One possibility is to sequentially chop off the amino acids to shorten the molecule and observe changes in structure-taste relationships.

Dipeptides

Mazur *et al.* (1970) recently reported the structure-taste relationships of 49 aspartic acid amides. However, the original dipeptide ester discovered (L-Asp-L-Phe Me ester) has good taste qualities with rather high receptor affinity. Unfortunately, amino acid imbalance, particularly excess of phenylalanine, is of concern in human nutrition. The question of interest is whether the dipeptide can somehow be incorporated into a small protein and still retain its excellent sweetening qualities. This would then solve the problem of amino acid imbalance, but the pH and temperature sensitivity must still be dealt with. Decline of activity with storage is also a serious factor.

Serendipity Berry

The active principle of this fruit is a protein (Morris and Cagan 1972), but identification of the active center has not yet been established. Its affinity is too high, so that it has a lingering effect. The sweet quality is not as good as that of the above dipeptide. This molecule does have the advantage of quick action and low threshold.

Nothing published about the structure of the serendipity berry sweetener indicates anything particularly special about its composition, and it contains no carbohydrates. The immediate question is whether the active center is a combination of amino acids. The amino acid sequence of this molecule would be of great interest, and further purification may decrease the lingering effect.

Proteins in other species

In the past few years several reports have indicated that chemoreceptors can respond to proteins. This came as a surprise to most researchers in the chemical senses, but it should not have. Protein interactions with other cell membranes are well-known.

Recently Gurin and Carr (1971) found that a marine snail responds to oyster juice. An active ingredient is a protein. After analysis and further tests, it was reported that human serum albumin is also a good stimulus molecule at a $10^{-9} M$ concentration.

Amino acids are known to be good stimuli for fish. L-serine is detected by salmon at very low concentrations (Hasler and Wisby 1951). Suzuki and Tucker (1971) and Caprio and Tucker (1973) studied the response of catfish to 23 amino acids, some of which stimulated at $10^{-8} M$ concentration. The economic importance of this research is obvious. The role of proteins and amino acids must be carefully examined, but they are often overlooked by flavor chemists.

Taste Mixtures

Mixtures of two sweeteners often taste either more or less sweet than predicted by algebraically adding their individual intensities of sweetness. The former case is often refered to as synergistic. Unfortunately there is no theoretical basis for such algebraic summations. Apparently this misconception is due to a lack of mathematical understanding. The function between sweetness intensity and concentration is not linear. A good psychophysical technique for measuring sweetness intensity is magnitude estimation. Moskowitz (1971) measured the relative sweetness of 43 sugars using this method and found the following relationships:

$$S = k\ c^n$$

where
   S = intensity of sweetness
   k = constant of proportionality
   c = molar concentration
   n = exponent

The value of n was approximately 1.3 for all the sugars studied and less than 1.00 for saccharin and cyclamate. The exponent n is a measure of the rate of growth of sweetness with concentration, and the value of k is a measure of the relative sweetness of sugars. A comparison of the intensity of a few sweeteners at $1.0M$ concentration is (Moskowitz 1970, 1971): sucrose, 3.20; fructose, 2.05; lactose, 1.45; glucose, 1.00; mannitol, 0.80; ribose, 0.68; sorbitol, 0.53; and glycerol, 0.37.

If solutions of two different concentrations of a sugar are mixed, and if their exponent is $n = 2$, then:

$$S_T = k(C_1 + C_2)^2$$
$$S_T = kC_1{}^2 + kC_2{}^2 + 2C_1C_2$$

On the other hand, if the individual sweetness were calculated and added algebraically:

$$S_1 = kC_1{}^2 \qquad S_2 = kC_2{}^2$$
$$S_T = kC_1{}^2 + kC_2{}^2$$

Thus the false assumption of simple summation leads to a prediction of total sweetness less than that actually obtained by the amount $2C_1C_2$. Similar, but not identical, results would follow if two different sweeteners were considered. Algebraic summation always leads to false synergism if the sweeteners have exponents greater than unity. Since the mathematical expression for taste mixtures is complex, a reasonable approximation is obtained if one of the sweeteners is matched to a concentration of the second sweetener, and then the sum of the two concentrations of the same sweetener used to obtain an equivalent sweetness. Many researchers have converted their sugars to glucose concentrations of equivalent sweetness.

In all the above considerations it was assumed that both sweeteners interact with the same receptor sites. Electrophysiological experiments indicate that this is not necessarily correct, and that many different sweet receptor sites exist. In a few cases true synergism may exist if the two sweeteners chosen do not compete for the same receptor site. The maximum sweetness that can be obtained, however, is less than double the intensity produced by the sweetest of the two chosen molecules in the mixture. Thus synergism of taste mixtures may help increase the sweetness of food, but sugar mixtures are not a substitute for a new potent sweetener to replace saccharin, for example.

CONCLUSION

Our knowledge of taste receptors and of structure-taste relationships is still rudimentary. However, taste and smell are the monitors for food intake which in turn determines nutrition. The implications for society are therefore great.

The immediate question of concern today is the sweetener. In the past most discoveries were accidental. Recently, a few natural sweeteners have been investigated and this should be continued. The literature suggests a number of plants that have been used for sweetening purposes (Uphof 1968). Research on dipeptide and protein stimuli of taste receptors should be greatly expanded. It is this area that shows greatest promise at the present time.

It should be stressed, however, that sugars are excellent foods, have good taste qualities, and have superior characteristics for incorporation into foods. The total individual consumption, however, is often too high. The best solution is not prohibition but temperance, with a variety of other sweeteners of high potency used in conjunction. Thus, not one but many new sweeteners are needed.

BIBLIOGRAPHY

BEIDLER, L. M. 1954. A theory of taste stimulation. J. Gen. Physiol. *38*, 133-139.

BEIDLER, L. M. 1967. Anion influences on taste receptor response. *In* Olfaction and Taste. II. T. Hayashi (Editor). Pergamon. Oxford, pp. 509-534.

BEIDLER, L. M. 1971. Taste receptor stimulation with salts and acids. *In* Handbook of Sensory Physiology, Vol. IV. Taste. L. M. Beidler (Editor). Springer Verlag, New York, pp. 200-221.

BEIDLER, L. M., and SMALLMAN, R. 1965. Renewal of cells within taste buds. J. Cell. Biol. *27*, 263-272.

BRADLEY, R. M. 1972. Duplexity of taste receptor loci. *In* Olfaction and Taste, IV. D. Schneider (Editor). Wissenschaftliche Verlagsgesellschaft, Stuttgart, pp. 271-279.

BRADLEY, R. M., and MISTRETTA, C. M. 1972. The morphological and functional development of fetal gustatory receptors. *In* Oral Physiology, N. Emmelin and Y. Zotterman (Editors). Pergamon, Oxford, pp. 239-253.

CAPRIO, J., and TUCKER, D. 1973. Amino acids as taste stimuli in the freshwater catfish, *Ictalurus Punctatus*. Fed. Proc. *32*, 328.

CUATRECASAS, P., and ILLIANO, G. 1971. Membrane sialic acid and the mechanism of insulin action in adipose tissue cells. J. Biol. Chem. *246*, 4938-4946.

DASTOLI, F. R. 1972. Fluorescence of the soluble "sweet-sensitive" protein complexes with sugars. Experientia *28*, 387.

DASTOLI, F. R., and PRICE, S. 1966. Sweet-sensitive protein from bovine taste buds: isolation and assay. Science *154*, 905-907.

DESNOO, K. 1937. Das trinkende Kind im Uterus. Mschr. Geburtsch. Gynak. *105*, 88-97. (German).

DESOR, J. A., MALLER, O., and TURNER, R. E. 1972. Taste in acceptance of sugars by human infants. Presented at Fourth Symposium on Oral Sensation and Perception: The Mouth of the Infant. N.I.H., Bethesda, Md.

GURIN, S., and CARR, W. E. 1971. Chemoreception in *Nassarius obsoletus:* The role of specific stimulatory proteins. Science. *174,* 293-295.

HASLER, A. D., and WISBY, W. J. 1951. Discrimination of stream odors by fishes and its relation to parent stream behavior. Am. Nat. *85,* 223-238.

KIER, L. B. 1972. A molecular theory of sweet taste. J. Pharm. Sci. *61,* 1394-1397.

KURIHARA, K. 1972. Personal communication. Tokyo, Japan.

KURIHARA, K., and BEIDLER, L. M. 1968. Taste-modifying protein from miracle fruit. Science *161,* 1241-1243.

KURIHARA, K., KURIHARA, Y., and BEIDLER, L. M. 1969. Isolation and mechanism of taste modifiers: taste-modifying protein and gymnemic acids. *In* Olfaction and Taste. III. C. Pfaffmann (Editor). Rockefeller University Press, New York, pp. 450-469.

LILEY, A. W. 1971. The pathophysiology of amniotic fluid abnormalities. *In* The Pathophysiology of Gestation, Vol. II. N. Assali (Editor). Academic Press, New York.

MAZUR, R. H., GOLDKAMP, A. H., JAMES, P. A., and SCHLATTER, J. M. 1970. Structure-taste relationships of aspartic acid amides. J. Med. Chem. *13,* 1217-1221.

MAZUR, R. H., SCHLATTER, J. M., and GOLDKAMP, A. H. 1969. Structure-taste relationships of some dipeptides. J. Am. Chem. Soc. *91,* 2684-2691.

MISTRETTA, C. M. 1971. Permeability of tongue epithelium and its relation to taste. Am. J. Physiol. *220,* 1162-1167.

MISTRETTA, C. M., and BEIDLER, L. M. 1967. Permeability of rat tongue epithelium. Physiologist *10,* 252.

MOSKOWITZ, H. R. 1970. Ratio scales of sugars sweetness. Percept. Psychophys. *7,* 315-320.

MOSKOWITZ, H. R. 1971. The sweetness and pleasantness of sugars. Am. J. Psycho. *84,* 387-406.

MORRIS, J. A., and CAGAN, R. H. 1972. Purification of monellin, the sweet principle of *Dioscoreophyllum cumminsii.* Biochem. Biophys. Acta *261,* 114-122.

SHALLENBERGER, R. S., and ACREE, T. E. 1967. Molecular theory of sweet taste. Nature (London) *216,* 480-482.

SHALLENBERGER, R. S. and ACREE, T. E. 1971. Chemical structure of compounds and their sweet and bitter taste. *In* Handbook of Sensory Physiology, Vol. IV. Taste. L. M. Beidler (Editor). Springer Verlag, New York, pp. 222-277.

SUZUKI, N. and TUCKER, D. 1971. Amino acids as olfactory stimuli in freshwater Catfish, *Ictalurus Catus* (Linn.). Comp. Biochem. Physiol. *40,* 399-404.

UPHOF, J. C. 1968. Dictionary of economic plants. Stechert-Hafner, New York.

ZAWALICH, W. S. 1972. Comparison of taste and pancreatic beta cell receptor systems. *In* Olfaction and Taste. Vol. IV. D. Schneider (Editor). Wissenschaftliche Verlagsgesellschaft, Stuttgart, pp. 280-286.

CHAPTER 3

Rose Marie Pangborn | # Sensory Perception of Sweetness

With the possible exception of salt, sugar appears in more different articles of the diet, either naturally or as an added ingredient, than any other food product. The universal liking of sweet substances by infants, children, adults, and many animals is well-documented. The human infant is exposed to sugar within minutes after birth through the lactose of mothers' milk, or the sucrose or glucose of sweetened formula milk. Newborn infants of both human and many other species show positive responses to sugars in essentially the same manner as do adults (Desor and Maller 1971; Grace and Russek 1969).

Sweetness is associated with emotive value judgments of what are thought to be psychologically, socially, and materially desirable qualities. In addition to gustatory references, dictionaries list synonyms for sweet and sweetness such as: pleasant, enjoyable, delightful, dulcet, clean, new, melodious, mellifluous, harmonious, redolent, aromatic, scented, winning, amiable, kind, gracious, charming, dear, precious, lovable, beloved, darling, and sweetheart. Perhaps Aykroyd (1967) is justified in stating that the word "sweet" and its derivatives have become the most overworked group of words in the English language.

On the other hand, not all usages of the word "sweet" are pleasant, e.g., sweet can sometimes refer to a person or situation which is cloying, sentimental, or unrealistic. "Sugary" may suggest excessive and offensive sweetness. "Sugared words" conveys a notion of beguilement, of an attempt on the part of the speaker to deceive by smooth speech. The expression "to sugar the pill", a reminder of the days when an important use of sugar was in disguising the unpalatable taste of medical nostrums, also carries the idea of deception. In most languages, however, "to sweeten" implies the enhancement of value.

Sweetness is a taste that is strongly associated with affection and reward. Few would doubt the enjoyment afforded to adults as well as children by the offering and receiving of gifts of confectionery. According to Lynch (1971), there seems to be an analogy between the relationship of sweet-tasting foods to overall food intake and leisure as it relates to the cultural activities and responsibilities required of the individual. Both sweet-tasting foods and leisure are socially permissible forms of relief from necessity and discipline.

23

A Brief History

The consumption of sweet substances in the form of fruits, berries, and honey predates written history. Anthropologists speculate that primitive man traveled considerable distances, tolerated painful bee stings, and fought off other animal species at the bee hive to obtain honey to satisfy his craving for sweets. One of the famous stone-age paintings found in caves in Southern Spain, depicts a man robbing a wild bee's nest of its honey dates back nearly 5,000 years. Bee-keeping is portrayed in tombs of the fifth dynasty in Egypt (2600 B.C.) and was practiced in the civilization of Sumer at about the same date (Aykroyd 1967). The highly prized sweet was considered worthy as an offering to the gods, forming the essential ingredient of the "soma", the drink of the gods of the Aryan invaders of India. The three most famous drinks of antiquity were all based on honey—the ambrosia and nectar of the ancient Greeks, and the mead of the Anglo Saxons and Vikings. The high regard for honey is still seen in primitive societies today, e.g., the Guayaki Indians of Paraguay use honey as their basic food and regard it so highly that their whole culture is based on it. For the aborigines of Australia, a "sugar bag" (honeycomb) is still a rare luxury, and they roam many miles of wild country in search of one (Cameron 1971).

Apparently sugar was unknown to the ancient Egyptians, Hebrews, Greeks, and Romans. The Promised Land flowed with milk and honey, but there is no mention of sugar in the Bible, the Talmud, or the Koran. Aykroyd (1967) points out that there are two references to "sweet cane" in Isaiah 43:24 and Jeremiah 6:20. Earliest reference to sugar is contained in the writings of officers of Alexander the Great during his campaign to India in 327 B.C.: "In India there is said to be a reed which yields honey without the help of bees, also that it yields an intoxicating drink even though the plant does not bear fruit" (van Hook 1949). Definite references to sugar are found in the Bower Scroll of 375 A.D., in which honey, sugarcane, sugar, and their decoctions are mentioned frequently.

Originally, sugar was refined from cane by the physicians of India and Arabia and used as medicine. Further proof of its origin is that the word for sugar in major languages is derived from Sanskrit *sákara*, and the word for candy from the Sanskrit *khanda*. Throughout ancient and medieval times, sugar was used primarily as a medicine. The renowned Galen extolled the virtues of sugar, which was recommended for disorders of the stomach, kidneys, and intestines. Gradually sugar became popular among the rich where it was prized as a luxury for beverages and confections. As early as the 7th century, sugar had become a commercial product. In the 14th century, increases in demand for sugar paralleled those for coffee, tea, and chocolate. Even today the four crops are interrelated. In the 16th century, sugarcane culture was introduced in Brazil and the West Indies, where one of its major uses was fermentation into rum (van Hook 1949). Despite the expanded production of sugar, its cost in 16-century England was equivalent to the cost of caviar today (Yudkin *et al.* 1971).

In 1747, crystals of sugar were discovered in the red beet by Marggraf, who suggested the possibility of extracting it on a commercial basis. Napoleon I furthered the development of sugar extraction from beets, because his "Continental System" forbade importation of items, including sugarcane, which might benefit the British financially (van Hook 1949).

Whereas sugar consumption in the United States and in the United Kingdom has increased about 10% in the past 30 yr, in several countries in Latin America, the Near East, and Africa, it has increased by from 200 to 300%, and in some cases much more. Among one group of Canadian Eskimos, sugar consumption rose from 26 lb in 1959 to 104 lb in 1967, while consumption among the rural Zulu population of South Africa increased from 6 lb in 1953 to 60 lb in 1964 (Yudkin 1972).

The relationship between sugar consumption and affluence has been summarized by Yudkin *et al.* (1971). As a country moves from poverty to affluence, there is a small rise in protein intake, a considerable rise in fat consumption, and a drop in starch consumption that is about the same as the rise in sugar. Much of the sucrose in the diets of the high-consuming countries is found in manufactured foods, such as candy, jams, jelly, soft drinks, ice cream, cakes, and cookies. Combining data from 8 countries, Yudkin *et al.* (1971) showed that as annual income rose from 38 to 460£ ($100 to 1150), daily per capita caloric intake rose from 2,060 to 3,200, sugar intake increasing from 35 to 139 gm. Data reported by the U.S. Department of Agriculture (1968) showed that poorer families (< $1,000 to $5,000/yr) consumed more household sugar than did higher-income families, who consumed must more of their sugar in the form of commercially prepared foods. This finding is consistent with a subsequent one reported by the U.S. Department of Agriculture (1969) that sugar in processed foods costs the consumer about twice as much as household sugar.

Per capita figures for the United States show that 102 lb of sucrose and 14 lb of corn sugar were consumed in 1970 (Briggs 1972). The highest consumption of sugar and sweets (excluding syrup, honey, and molasses) is by children between 12 and 14 yr of age: 49 gm/day for boys and 43 gm/day for girls (USDA 1969). Between ages 20 and 34, the daily intake is 37 gm for men and 31 gm for women. Over 65 yr of age, the figure decreased to 29 gm for women, but increased up to 40 gm for men. Depending upon age group, socio-economic level, and country of origin, consumption of sucrose can amount to between 1/6 and 1/3 of the daily caloric intake.

Today, the British eat more sweets than any other people and consume more sugar than most. Britain is the world's largest exporter of sweets, meeting 1/3 of the world demand. Lynch (1971) offers three reasons why sugar consumption in Britain exceeds that of the U.S.: (1) Americans are more concerned with their health and appearance; (2) Britain consumes considerably less fruit than other countries, preferring instead the cheaper cakes and puddings; and (3) what is known as "plain English cooking" may make the palatability of sweet-tasting food a necessary relief in the daily food patterns.

Watson (1971) attempts to explain the high intake of sweets in Britain as due to the continuation of behavior patterns developed as part of the cultural system rather than to some enhanced physiological need for sucrose. The satisfaction of social needs, when added to the pleasurable experience of sweetness, strengthens the preference for sweet food. Sweets and confections are used to enhance personal relationships, utilizing pleasurable experience which accompanies sweet sensation. Sweets are used as gifts between friends, and as rewards for children, creating feelings of approval, friendship, relaxation, and a sense of belonging. Factors such as these could account for some people resorting to sweet-eating in times of stress.

WHY SUGAR?

Many people obviously consider sugar essential as far as palatability of their food is concerned; however, sugar is not an essential nutrient. Why, then, is sugar so popular? In the author's opinion, there are four main reasons for the high consumption of sugar, in decreasing order of priority:

(1) *Sugar is sweet.* Its sweet taste has no undesirable overtones of bitterness, sourness, or saltiness. Man derives much pleasure and oral satisfaction from the sensation of sweetness. Sweet foods are attractive far beyond their value in relieving hunger, and sweet drinks are consumed far in excess of their value in relieving thirst (Cabanac 1971). It has long been known that the sweetness of sucrose can

serve as a very effective reward for many domestic and laboratory animals.

(2) *Sugar is economical.* Whereas wheat, potatoes, and maize yield 1 million, 2 million, and 3 million Calories per acre, respectively, cane sugar or beet sugar will yield 7.5 million Calories per acre. Because beets and sugarcane are extremely efficient converters of solar energy into Calories, it has been suggested that expansion of sugar production would provide one means of increasing world food supplies and conserving space needed for necessary, but less efficient, production of high-quality protein (Hockett 1955). In most parts of the world, the price of sugar has fallen steadily during recent decades, while that of other common foods—bread, rice, potatoes, and especially protein foods—has risen. In the 14th century, a pound of sugar cost the equivalent of a workman's weekly wage; today a pound of sugar costs only about 15 cents (Nordsiek 1972).

(3) *Sugar is versatile.* It has been incorporated into hundreds of foods and beverages where it serves several functions besides providing sweetness (ACS 1955). Its use in different candies depends upon its bulk, on its ability to exist in either crystalline or noncrystalline forms, on its solubility in water, and on its change of color and flavor when heated. So far the only way to produce a crystal-clear, hard and brittle sweet is with sucrose (van Eijk 1972). In the preparation of jams and jelly, sugar combines with acid and pectin to thicken or solidify the products, thereby preserving the fruits or berries. Sugar also is an effective preservative in sweetened condensed milk, where it inhibits mold and bacterial growth as a result of the osmotic pressure of solutions in high concentrations.

Sugars are important in development of color in the curing of meats, such as bacon and ham. Sucrose improves color by establishing reducing conditions, by tending to prevent oxidation of ferrohemoglobin to ferrihemoglobin during storage, and by helping to conserve the meat during curing by its protein-sparing action. In sausages and other processed meats, small additions of sucrose produce a better blend of tastes.

Sugar is essential in baked goods as the source from which the yeast produces the CO_2 necessary to the rising and leavening action. The desirable light-brown colors formed on bakery products are due to the ability of sucrose to caramelize upon heat treatment. Not only does sucrose contribute to the flavor and color of breads, cakes, and biscuits, and other bakery products, but also to their texture.

In many beverages, sucrose not only provides sweetness but also increases viscosity and thereby the desirable body of the product. Sucrose has many uses as a seasoner or condiment where low concentrations enhance natural flavors, reduce off-flavors, decrease apparent bitterness and sourness, and improve body and texture (Sjöström and Cairncross 1955).

(4) *Sugar is convenient.* It dissolves quickly in water to give a colorless solution. It is easy to transport, to store, to keep clean, and has a relatively small bulk in proportion to its high caloric value. For example, if an adolescent boy requires 4,000 Calories per day, it would be difficult for him to eat enough fresh fruits and vegetables and protein foods to meet this need without distending his stomach as well as his father's finances.

THE TASTE OF SUGAR

Despite the ever-increasing proliferation of literature on sweetness and sweeteners, Moncrieff (1970) is dismayed that the questions for which there were no answers in 1940 remain unanswered today, namely: (1) What is the function of the sweet taste? Why do we like sweetness so much? Is it associated with some physiological need? (2) When phylogenetically did the sweet taste appear and why? (3) Do the receptors for the sweet taste respond uniquely to it, or do they respond to all tastes? Are the receptors for other tastes sensitive to a sweet compound? (4) Is there a consistent relationship between sweetness and any physical or chemical property? and (5) Is there any significance in the intense sweetness of substances such as saccharin, or is it just fortuitous?

In this chapter, highlights from research on the sweet taste will be cited to draw attention to the major experimental variables that influence sweetness: (1) variations within the experimental subject; (2) variations inherent in the physical or chemical stimulus; and (3) variations contributed by the test procedures and the external environment. No attempt will be made herein to present a detailed summary of the vast, diverse literature on the sweet taste. The divergent and sometimes contradictory results reported in the early as well as contemporary literature have been attributed by Meiselman (1972) to: (1) poorly defined goals and poorly standardized procedures; (2) unknown or uncontrollable variables; (3) failure to select experimental subjects according to defined criteria besides availability; and (4) the questionable relevance of taste research to realistic situations involving gustatory stimuli, e.g., food and beverage consumption. The author will attempt to resist the temptation to extrapolate results from laboratory studies to dinner-table conditions, and advises the reader to do likewise.

Function of the Sweet Taste

According to Lynch (1971) and others, the primary function of taste sensitivity lies in determining which parts of the environment contain substances suitable as food and which do not. Undoubtedly gustation and olfaction are important in the recognition of food and appreciation of food quality, but the influence of the cultural norm on food acceptance should not be overlooked. For example, relative to "nonfood" items, we note in several societies the practice of geophagia (clay-eating) and entomophagy (insect-eating). These items may be nonfoods to most Americans, yet many populations around the world relish locusts, caterpillars, termites, beetles, and other insects (Holt 1885; Essig 1934). Furthermore, people such as the Pomo, Lake Miwok, and Central Wintun Indians of California mixed red earth with their acorn meal, while the Plains and Northern Miwok sometimes mixed ashes with acorn dough (Gifford 1971). Therefore, one man's pica could be another man's dinner.

Sensations of sweetness and saltiness are definitely associated with foods and beverages, but sugar played a very small part in nutrition until man learned to cultivate it in abundance. As pointed out by Renner (1944), there was little need for sugar to be registered by a special receptor, as it formed the smallest part of man's carbohydrate consumption until the 19th century. The human had a much greater need for an oral receptor registering starch, which formed the main part of the carbohydrate consumption. Renner believes there is no nutritional usefulness for the sweet taste, because if there were, the receptors would be located in the intestines where starch is converted to the sweet compound, maltose. Therefore, the location of the sweet receptors in the mouth was of little nutritional significance, until the relatively recent abundance of ready-made sugar.

Despite the non-teleological placement of the receptors for the sweet taste, primitive man had a sweet tooth and satisfied his craving primarily by eating fruit. In the process, he obtained nutrients, such as ascorbic acid which he needed. (According to Aykroyd (1967), one unquestionably beneficial effect of sugar on the dietary habits of Britain was that it encouraged the consumption of fruit, as many fruits taste better when eaten with sugar.) Modern man, on the other hand, can satisfy his desire for sweetness by consuming foods and beverages which provide little or no nutrition besides calories. There is no physiological need for sucrose in man or animals. Sheffield and Roby (1950) observed that rats drank nonnutritive saccharin solutions under prolonged exposure to them, showing no sign of extinction such as might be expected if the preference for saccharin or "sweetness" were acquired through past association with nourishment. This demonstrates that there is little or no relationship between preferences and nutritive value in normal subjects. Yudkin (1972) believes that it is the dissociation of palatability from nutritional value that is the major cause of the current "malnutrition of affluence".

Sweet Quality

Sweetness is associated with aliphatic hydroxyl compounds, but sweet-tasting compounds include such diverse molecules as sucrose, dulcin, saccharin, glycine, chloroform, lead acetate, and beryllium chloride. Many glycols are sweet, several α-amino acids are sweet, a few aldehydes and ketones are sweet, as are many esters. Some sweet compounds are toxic, such as the n-propyl derivative of 4-alkoxy-3-aminonitrobenzene, which is about 5,000 times sweeter than sucrose (Amerine et al. 1965).

Several compounds are sweet at low concentrations but become bitter with increasing concentrations. The sweetener saccharin has been noted for its concommitant bitterness, which some consumers find objectionable (Helgren et al. 1955). Cameron (1947) reported secondary taste sensations for sugars such as: glucose = sweet plus bitter or sour; lactose = sweet plus salty; glycerol = bitter-sweet; galactose = sweet plus leaden or woody; mannitol = sweet plus flat or woody; glycine = sweet plus bitter or sour or sharp; and DL-alanine = sweet plus sour or sharp or ether-like. As stated by Cameron (1947), " ... the tasters with lesser delicacy of perception usually did not perceive an additional taste and were not misled by it, but those with very delicate taste perception were occasionally so influenced by the extra taste sensations they perceived, that they sometimes had difficulty in comparing sweetnesses, and in one or two instances were incapable of doing so." Subjects tested by Dahlberg and Penczek (1941) observed that the sweetness of sucrose was perceived quickly, promptly reached a maximum intensity, and then decreased. The sweetness of glucose, on the other hand, stimulated the taste receptors more slowly and reached a maximum intensity later.

A renewed interest in multidimensional scaling of gustatory qualities is represented in the article by Schiffman and Erickson (1971), which describes judgments of similarities and semantic differential ratings of 19 taste stimuli. Both physical and psychophysical continua were designated as useful dimensions represented by three criteria: relative goodness or badness of the tastes, molecular weight, and departure from pH 7. The semantic differential descriptions for the sweet compounds included: Sucrose—"This is a sweet stimulus which is good, pleasant, smooth, flavorous, and moderately soft, although it is neither particularly refreshing nor sensual. It is a simple, foodlike taste which in general develops rapidly and remains constant." Glucose received the same description. Saccharin—"This is a sweet, flavorous taste which is consistently rated as complex. Two subjects found it to have a bitter taste, with this taste developing slowly and changing over time. Two other subjects rated it exactly opposite, i.e., a nonbitter, fast-developing taste which remains constant. The complexity of this stimulus leads to considerable individual variation in judgment, with a wide range of rating given for the goodness, poisonousness, and foodlike qualities of the taste. Saccharin has no burning or tingling components." One cannot help but wonder whether descriptions from subjects of different ages, educational levels, cultural back-grounds, or speaking different languages, might produce similar or completely different subjective statements.

Using magnitude estimation with a series of sugars, only, Moskowitz (1972A) measured "flavor" differences between pairs of equally sweet concentrations of sucrose, fructose, glucose, lactose, maltose, sorbose, galactose, xylose, arabinose, glycerol, and sorbitol. Two dimensions were apparent in the perceptual space, i.e., viscosity-fluidity, and sweet versus sweet plus side tastes.

Recent cross-adaptation studies (Meiselman 1968; McBurney 1972) suggest that a single receptor mechanism may account for the sweet quality. In cross-adaptation tests, it is established whether prolonged taste stimulation with resulting sensory decrement with one compound depresses responsiveness to another test compound, to determine whether the two share common loci of stimulation on the gustatory receptors. McBurney (1972) compared an oral prerinse of water with that of $0.32M$ sucrose and $0.01M$ saccharin. Sucrose adaptation reduced the apparent sweetness of 11 sweet compounds, as did saccharin, but to a considerably lesser degree. The sweetness of sucrose was reduced about 80% by a sucrose rinse, while saccharin adaptation reduced the sweetness of saccharin by 55%.

The sweetness of a compound is related to its molecular configuration and other chemical properties (Birch *et al.* 1970; Birch and Lee 1971; Shallenberger and Acree 1971). Shallenberger proposes that sweet-tasting compounds possess AH and B units, where H is an acidic proton and B is an electronegative center, with a distance of about 3 Å between the two. Dzendolet (1968) suggests that sweet-tasting compounds are proton acceptors, with a model which accounts for the sweetness of dilute salts, as well as for common sweeteners.

Sweetness also is related to solubility characteristics. Solutions of 40% sucrose or more taste sweeter than 100% granulated sugar, which in turn, tastes sweeter than a lump of sugar. The lump dissolves on the tongue much more slowly than the small crystals, producing a lesser sweetness intensity. With solutions of high concentration, the taster is not able to dilute the sweetness sufficiently with saliva and hence finds the solution sweeter than 100% sucrose. (The same observation can be made with rock salt versus a salt solution.) Renner (1944) had subjects suck sweets and expectorate the saliva instead of swallowing it. He calculated that the saliva had only 16% sucrose, suggesting that the pleasantness of sweetness lies far below that of a concentrated solution. Pariser (1961) found that the sugar concentration in the saliva of people sucking hard candy varied from 10 to 25%, concluding that these sucrose concentrations apparently represented the highest preference levels for each person for the specific candy. It appears that the enjoyment of sweetness can be reached only by slow consumption.

SWEETNESS RESPONSES

Pleasantness of Sweet Substances

The affective dimension, or pleasantness of sweeteners has received surprisingly little systematic investigation until recently. Early work by Engel (1928) showed that sucrose was pleasant at intermediate, but unpleasant at high concentrations. In a choice from an array of eight solutions, Chappell (1953) noted that a solution of 25% sucrose was the most acceptable, with a descending sugar preference order of sucrose, lactose, glucose, and maltose. In more extensive testing, Moskowitz (1971) observed that, unlike the sweetness intensity function, pleasantness of 43 sugars were not monotonic with concentration, departing from linearity at the extremes of concentration. Pleasantness as a function of sweetness was roughly linear in log-log coordinates, with a slope between 0.3 and 0.5.

Inspection of responses from individual subjects revealed bimodal hedonic distributions as a function of sucrose concentration (Engen *et al.* 1961). Increasing the concentration series, as illustrated in Fig. 3.1, Pangborn (1970) noted trimodal distributions of hedonic ratings for both sucrose and sodium chloride. Some subjects demonstrated increased degree of liking, some expressed increased dislike, and a third group liked increasing concentrations up to a maximum, followed by a distinct reduction. If the concentrations had been extended, subjects displaying continued increases in liking would

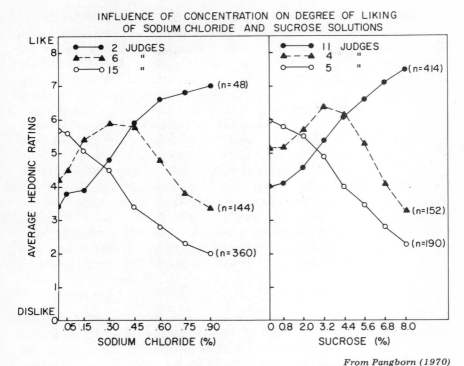

INFLUENCE OF CONCENTRATION ON DEGREE OF LIKING
OF SODIUM CHLORIDE AND SUCROSE SOLUTIONS

From Pangborn (1970)

FIG. 3.1. HEDONIC RESPONSES OF EXPERIENCED JUDGES TO INCREASING
CONCENTRATIONS OF SODIUM CHLORIDE AND SUCROSE IN SOLUTION

There is no correspondence between judges who liked or disliked sodium chloride and those
who liked or disliked sucrose.

probably have reached a plateau and then dropped. Note in Fig. 3.1
that responses to the distilled water blank differed among the three
groups, indicating they started from different baselines. One could
question the practicality of having trained subjects measure
acceptance of test solutions under artificial laboratory conditions,
i.e., what is there to like or dislike about a set of coded sugar
solutions at mid-morning, served in an antiseptic laboratory under
red illumination? Like and dislike for sweetness in lemonade or wine
might be less ambiguous than for model systems which are seldom
consumed, and for which multiple frames of reference might exist. It
would be considerably more informative if data such as the foregoing
were available, showing the influence of the subject's past
experience, expectation, sensitivity, and knowledge of the test
procedures on degree of liking for sweetness in a variety of foods and
beverages.

Preferences among solutions of water, 5% sucrose, 30% sucrose, 0.5% sodium chloride, and saccharin equivalent in sweetness to 30% sucrose were measured for 100 patients in various stages of hypoglycemia, covering all blood glucose levels from fasting to that of incipient hypoglycemic coma (Mayer-Gross and Walker 1946). The normal dislike for the excessive sweetness of a 30% sucrose solution was reversed into a distinct preference for the liquid when the blood glucose level dropped below 50 ml/100 ml. The solution of 5% sucrose generally was considered sweet above, and tasteless below blood glucose levels of 50 ml/100. Although the taste of the saccharin solution was described as generally pleasant, the solution was seldom preferred, despite the low blood glucose levels. In contrast, Pfaffmann (1957) severely lowered the blood sugar levels of hamsters by injection of insulin, and found no differential enhancement or suppression of sensitivity to sugar compared with that to sodium chloride. Furthermore, lowering the blood sugar did not appear to affect sensitivity of the peripheral gustatory receptors.

Sweetness and Concentration

Taste thresholds, i.e., the minimum concentration that can be perceived, have been measured with a variety of psychophysical techniques for many sweeteners, with an equal diversity of results. In an extensive literature review, Pfaffmann (1959) summarized threshold data for sucrose (0.171 to 0.548%) and glucose (0.721 to 1.621%), as well as for other sweet compounds. Although threshold data are reported frequently, no specific emphasis will be placed on sweetness thresholds in the present review. The author believes that thresholds are convenient, statistically determined endpoints which have informational value of a relative nature, and have application in terms of detection of off-tastes and off-flavors. However, thresholds fluctuate widely from trial to trial, from person to person, and with different test methods. More important, threshold values may not be predictive of relative intensities at the higher concentrations (Pangborn 1971).

It is well known that the relative sweetness of sugars varies with concentration, for example, at concentrations of 10, 25, and 40% sucrose, glucose was 65% as sweet, 71% as sweet, and 83% as sweet, respectively (Lichtenstein 1948). Using ratio scaling, Moskowitz (1970A) reported that the relative sweetness of 15 sugars (all except mannose) remained constant across concentrations, and that the sweetness increased according to a power function, expressed as: Sweetness = $k \times$ concentration$^{1.3}$, where the intercept, k, is a measure of relative sweetness. However, in comparisons of sucrose with saccharin, cyclamate, and mixtures of the latter two, relative sweetness was observed to change continuously with concentration (Moskowitz 1970B). In general, the artificial sweeteners increased in sweetness less rapidly than did sucrose, and became somewhat bitter at higher concentrations.

In addition to concentration, the medium of dispersion can influence the relative taste intensity of two compounds. For example, although fructose is sweeter than sucrose at all concentrations in distilled water, the two sugars were of equivalent sweetness when compared in peach and pear nectars (Pangborn 1963).

Taste Mixtures

The effect of one taste on another has been of considerable interest because foods and other oral stimuli contain a mixture of tastes and flavors. In most instances, one taste will reduce the perceived intensity of a second taste (Pangborn 1960). In particular, acids and sugars have mutually masking effects, as do salt and sugar (Pangborn 1961, 1962). A noted exception was the observation of a bimodal response to the influence of acid on apparent saltiness.

In a recent investigation by Moskowitz (1972B), not only was the sweetness of glucose and fructose reduced by additions of sodium chloride, citric acid, or quinine sulfate, but qualitative changes occurred as well. Mixtures of sweet and salty stimuli were considered to clash and be "unblended", in contrast to mixtures of sweet with either sour or bitter tastes. Subsequently, Moskowitz (1972C) reported on the interrelationship of sweetness with the cost of mixtures of sucrose, glucose, fructose, sorbitol, sodium saccharin and sodium cyclamate.

Several researchers have found that the sweetness of sucrose solutions are enhanced by addition of ethyl alcohol (Berg *et al.* 1955; Hellekant 1967; Martin and Pangborn 1970). Added thickening agents also can modify sweetness; for example, Stone and Oliver (1966) reported that the intensity of 1, 2, 5 and 10% sucrose in solution was enhanced somewhat by the viscosity imparted by 2% cornstarch or by 1% gum tragacanth. Using sucrose in combination with cornstarch, guar, and carboxymethylcellulose (CMC), Vaisey *et al.* (1969) measured the rates of sweetness recognition, matching of equisweetness in different gums, apparent sweetness intensities, and ranking of sweetness within each gum. The cellulose gum, with less drop in viscosity with increasing rates of shear, tended to mask sweetness perception throughout. According to Moskowitz and Arabie (1970), increases in viscosity imparted by CMC generally decreased the taste intensity of glucose solutions. Similar conclusions were reported in a subsequent publication, where increasing viscosity was stated to reduce the sweetness of both sucrose and sodium saccharin (Arabie and Moskowitz 1971). In contrast, Pangborn *et al.* (1973) found that the presence of hydrocolloid thickeners generally depressed the sweetness of sucrose, while significantly enhancing the sweetness of saccharin (See Fig. 3.2). The discrepancies in reported results may be attributed to the fact that different concentration ranges were tested by different investigators.

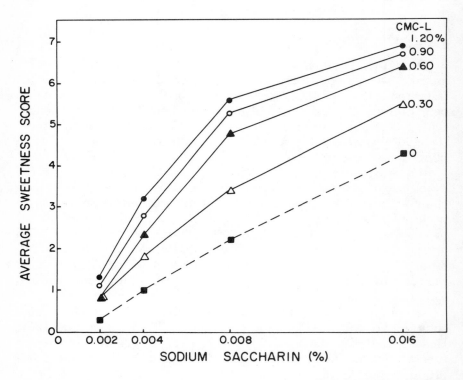

From Pangborn et al. (1973)

FIG. 3.2. SWEETNESS INTENSITY RESPONSES OF EXPERIENCED JUDGES TO INCREASING CONCENTRATIONS OF SODIUM SACCHARIN, WITH AND WITHOUT CARBOXMETHYLCELLULOSE-L, AN EDIBLE FOOD THICKENER

Points connected by solid lines are averages of 32 separate evaluations, whereas those connected by the dotted line are based on 128 evaluations.

Undoubtedly the sweetness of many complex foods and beverages can be modified by changes in the physical and chemical composition of the mixture. Much experimental work remains to be done before there is a complete understanding of sweetness quality and intensity in complex food systems.

Sex and Age Differences in Sweetness Responses

The suggestion that hormone differences may influence preferences for carbohydrates has been advanced by several investigators. Watson (1964) found that when male rats were presented with a choice of diets varying only in the type of carbohydrate, they selected a diet with more dextrin and less sucrose than did females. Shim (1968) reported that when protein was increased at the expense of carbohydrates, rats preferred a higher proportion of sucrose to starch as the carbohydrate. On the other hand, when protein was increased at the expense of fat, this relationship was shown mainly on the first day, and was much reduced thereafter. Possibly, as the protein changed, the taste-texture sensations changed in such a way that sucrose made the diet more acceptable. An interesting related finding was that men and post-menopausal women responded to dietary sucrose with an increase in serum triglycerides— a response that was absent in pre-menopausal women (MacDonald 1971; Coltart 1969). In rats, ingestion of sucrose caused an increase in liver triglycerides in males but not in females (Morton and Horner 1961).

Valenstein *et al.* (1967) found that when rats were given their choice between water and a 3% glucose solution (barely sweet to humans), both sexes preferred the glucose solution; the females consumed more of a 0.25% solution of saccharin (sickeningly sweet for humans) than of the 3% glucose solution, but the males switched to glucose by the third or fourth day of exposure. In contrast, the females consumed greater amounts of saccharin and did not switch. During a third experiment, the female rats continued to exhibit a clear preference for increasing concentrations of saccharin solutions (up to 1.05%), compared to tap water, while the males consumed equal amounts of tap water and saccharin solutions.

Quantitative changes in taste perception with age have been studied by several investigators. Richter and Campbell (1940) found that the taste threshold for sucrose in people between 52 and 85 yr of age was three times higher than that of the 15 to 19 yr age group. Cooper *et al.* (1959) reported a sharp decline in sensitivity for all 4 basic taste qualities after the late 50 years of age. No differences were observed between the two sexes or between cigarette smokers and nonsmokers. Byrd and Gertman (1959) concluded from their study of 40 subjects between the ages of 60 and 90 yr that there is no evidence for a decrease in taste sensitivity with age, but that a few selected people show marked impairment, which they attributed to disease rather than to senescence.

Glanville *et al.* (1964) tested subjects ranging from 3 to 55 yr of age and concluded that taste thresholds increase with age, the sensitivity of males deteriorating more rapidly than that of females. In contrast, Kaplan *et al.* (1965) did not observe a significant drop in taste sensitivity in subjects between 15 and 55 yr, except among smokers. Using electrical stimulation of the tip of the tongue, Hughes (1969) tested 29 students between 21 and 30, and 88 patients between 62 and 96. Very few of the young subjects needed more than 10 μA stimulation before a taste sensation was obtained, whereas the elderly showed a progressive elevation in the taste threshold with age, 7 requiring over 100 μA to evoke a taste sensation. These results parallel the anatomical observations of a decrease in the number of taste buds with age; for example, Arey *et al.* (1935) reported that children averaged 248 taste buds per papilla while a group of people between 74 and 85 yr averaged only 88 buds, over half of the buds in each papilla being atrophied.

Hunger, Obesity, and Sweetness Responses

To determine whether overweight people prefer sweeter desserts, degree of liking for ice cream and for canned fruit was measured among 12,505 consumers (Pangborn and Simone 1958). There was a tendency for the sweetest samples to be scored highest by overweight and lowest by underweight consumers; however, preference between two samples differing in sweetness was not influenced by body size. The tendency for obese individuals to indicate a high acceptance for all samples suggest that overeating and subsequent obesity may result from a greater liking for food in general rather than a particular preference for sweeter foods (Pangborn 1967).

Moore *et al.* (1965) measured the sensitivity of 7 subjects to sucrose solutions before and after a lunch and a no-lunch condition. Sensitivity was higher in the afternoon than in the morning and was independent of whether or not lunch was consumed. These data indicated that diurnal effects, rather than eating *per se*, influence sensitivity to sucrose. Pangborn (1959) found that sucrose thresholds (detection and recognition) are dramatically reduced with training, with no significant differences between fasting and non-fasting conditions in 7 out of 8 subjects.

Using 22 fasted subjects, Cabąnac *et al.* (1968) reported that stimulation with 20% sucrose solutions at 3-min intervals for 2 hr became unpleasant only when sucrose was placed in the stomach, either by swallowing or by intubation. This transformation of a pleasant sweet sensation into an unpleasant one was not observed after mechanical or osmotic stimulation of the stomach. Furthermore, no correlation was found between pleasantness of the sugar solution and blood glucose levels. These investigators concluded that the sensation of sweetness is composed of a discriminative element derived from oral stimulation and an affective element derived from internal stimulation.

Cabanac and Duclaux (1970) recorded the pleasantness of sucrose solutions by 5 male and 10 female fasted, nondiabetic obese patients, 6 nonobese females serving as controls. Solutions of 20% sucrose were considered pleasant by both groups, but became unpleasant to the controls only, shortly after ingestion of 50 gm of glucose in 200 ml of water. No correlation was found between sweetness responses and blood glucose or insulin. The authors concluded that obese people generally are unaware of internal signals for control of food intake.

Subsequently these investigators reported that increasing concentrations of sucrose (2.5 to 40%), which became unpleasant to normal subjects after injection of 50 gm glucose in 200 ml water, were unchanged in degree of liking after the same subjects had purposely lost 7 to 12 lb (Cabanac 1971; Cabanac *et al.* 1971). After returning to their initial body weight, the subjects gave taste responses almost identical with those at the beginning of the experiment. The researchers propose the existence of a biological "ponderostat" and conclude that the pleasantness-unpleasantness of food-related stimuli depends on a certain body weight set point. Based on his experiments, Cabanac believes that a stimulus (gustatory, olfactory, or thermal) feels pleasant or unpleasant depending upon its usefulness as determined by internal signals, a phenomenon which he calls "alliesthesia".

Personality and Sweetness Perception

In his review of the social aspects of sweetness preferences, Watson (1971) asks the provocative question, Is the liking for carbohydrates in general, and sugar in particular, mediated by social and psychological factors, by biochemical and physiological mechanisms, or a combination of the two? He cited data reported by Lát (1959) and Franková (1966) showing that rats selecting a relatively high protein diet demonstrated less spontaneous activity and less exploratory behavior, and developed conditioned responses more slowly, than did rats selecting a relatively high carbohydrate diet. The high-carbohydrate eaters could be characterized by an "excitatory" nervous system in contrast to the "inhibitory" one exhibited by the high-protein eaters. Feeding a particular type of diet was shown to produce changes in the "excitatory"/"inhibitory" level of the animal.

Watson (1971) commented on a small, but significant, correlation between sucrose intake in beverages and "extraversion", as measured in two groups of male students, using the Eysenck Personality Inventory. The observed absence of such a relationship in the comparable group of female students was interpreted as due to possible restriction of sucrose occasioned by dieting or fashion purposes. Watson expressed no doubt that other aspects of personality could be shown to be closely related to sucrose preferences, or to total carbohydrate preferences, in general.

It has been observed that rats fed in isolation from each other ate less than when fed in groups (Watson 1964)—a phenomenon known to occur among humans, too. Rats housed singly preferred more sucrose to dextrin than did animals housed in pairs. Watson interpreted his results as deriving from "stress", i.e., since rats are group-living animals, the isolated state was more "stressful" than the group state. Expressed in human terms, the sweetness provided a pleasant feeling that allayed anxiety and increased the feeling of well-being.

CONCLUDING STATEMENTS

The universal desire for the pleasantness derived from the sweetness of sugars continues to manifest itself in world-wide sugar consumption figures; industrialized areas, such as the United Kingdom and the United States, seem to have leveled off at an annual per capita intake of approximately 50 kg, while developing countries are rapidly approaching the same figure. In addition to its sweetness, sugar is attractive to the consumer because of its relative low cost, its convenience and its versatility. To the dismay of dentists, nutritionists, and physicians such as Yudkin (1972), our society continues to encourage children to indulge their sweet tooth, establishing food habits which persist into adulthood. A better understanding of the mechanism and function of the pleasantness of the sweet sensation could contribute toward alternative behavioral patterns with fewer undesirable side effects.

The sweetness of sugars has been investigated by researchers from a wide range of disciplines—food science and technology, stereo-chemistry, biochemistry, nutrition, neurophysiology, psychology, and sociology, to name a few. Whereas a great deal is known about several physical, chemical, physiological, and environmental variables that influence recognition, discrimination, and preference for sweetness, there are great gaps in our knowledge. Many significant advances have been made recently, as described herein and in the other chapters in this book. As a food scientist, I can visualize that considerable usefulness would derive from multidisciplinary studies of factors influencing the sweetness of processed foods and beverages, thereby paving the way for the substitution of noncariogenic, noncaloric sweetness which is much in demand in affluent societies.

BIBLIOGRAPHY

AMERICAN CHEMICAL SOCIETY, 1955. Use of Sugars and Other Carbohydrates in the Food Industry. Advances in Chemistry Series No. 12, ACS, Washington, D.C.

AMERINE, M. A., PANGBORN, R. M., and ROESSLER, E. B. 1965. Principles of Sensory Evaluation of Foods. Academic Press, New York.

ARABIE, P., and MOSKOWITZ, H. W. 1971. The effects of viscosity upon perceived sweetness. Percept. and Psychophys. 9, 410-412.

AREY, L. B., TREMAIN, M. J., and MONZINGO, F. L. 1935. The numerical and topographical relation of taste buds to human circumvallate papillae throughout the life span. Anat. Rec. 64, 9-26.

AYKROYD, W. R. 1967. The Story of Sugar, Quadrangle Books, Chicago.

BERG, H. W., FILIPELLO, F., HINREINER, E., and WEBB, A. D. 1955. Evaluation of thresholds and minimum difference concentrations for various constituents of wines. II. Sweetness: the effect of ethyl alcohol, organic acids and tannin. Food Technol. 9, 138-140.

BIRCH, G. G., and LEE, C. K. 1971. The chemical basis of sweetness in model sugars. In Sweetness and Sweeteners, G. G. Birch, L. F. Green, and C. B. Coulson (Editors). Applied Science Publishers, Ltd., London.

BIRCH, G. G., LEE, C. K., and ROLFE, E. J. 1970. Organoleptic effect in sugar structures. J. Sci. Fd. Agr. 21, 650-653.

BRIGGS, G. M. 1972. Consumer and food industry equally responsible for poor nutrition. Food Prod. Develop. 6, (7) 11.

BYRD, E., and GERTMAN, S. 1959. Taste sensitivity in ageing persons. Geriatrics 14, 381-384.

CABANAC, M. 1971. Physiological role of pleasure. Science 173, 1103-1107.

CABANAC, M., and DUCLAUX, R. 1970. Obesity: absence of satiety aversion to sucrose. Science 168, 496-497.

CABANAC, M., MINAIRE, Y., and ADAIR, E. R. 1968. Influence of internal factors on the pleasantness of a gustative sweet sensation. Communs. Behav. Biol. Part A, 1, 77-82.

CABANAC, M., DUCLAUX, R., and SPECTOR, N. H. 1971. Sensory feedback in regulation of body weight: is there a ponderostat? Nature 229, 125-127.

CAMERON, A. G. 1971. In search of sweetness. In Food Facts and Fallacies. Faber and Faber, Ltd., London.

CAMERON, A. T. 1947. The Taste Sense and the Relative Sweetness of Sugars and Other Sweet Substances. Scientific Report Series No. 9, Sugar Research Foundation, Inc., New York.

CHAPPELL, G. 1953. Flavour assessment of sugar solutions. J. Sci. Fd. Agr. 4, 346-350.

COLTART, T. M. 1969. Cited by A. E. Bender and K. B. Damji, 1971. Some effects of dietary sucrose. In Sugar, J. Yudkin, J. Edelman, and L. Hough (Editors). Butterworths, London.

COOPER, R. M., BILASH, I., and ZUBEK, J. P. 1959. The effect of age on taste sensitivity. J. Geront. *14*, 56-58.

DAHLBERG, A. C., and PENCZEK, E. S. 1941. The relative sweetness of sugars as affected by concentration. New York Agric. Exp. Sta. Tech. Bull. No. 258, Ithaca, N.Y.

DESOR, J., and MALLER, O. 1971. *Cited by* J.,A. Morris, 1972. Sweet taste, basic research with practical applications. Mfg. Confect. *52* (7), 38-39.

DZENDOLET, E. 1968. A structure common to sweet-evoking compounds. Percept. and Psychophys. *3*, 65-68.

ENGEL, R. 1928. Experiments examining the dependency of like and dislike of taste as a function of stimulus strength. Archiv. gesamte Psychologie *64*, 1-36. (German).

ENGEN, T., McBURNEY, D. H., and PFAFFMANN, C. 1961. The sensory and motivating properties of the sense of taste. *Cited by* C. Pfaffmann in M.,R. Jones (Editor). Nebraska Symposium on Motivation, 1961. Univ. of Nebraska Press, Lincoln, Neb.

ESSIG, E. O. 1934. The value of insects to the California Indians. Sci. Mon. *38*, 181-186.

FRANKOVÁ, S. 1966. *Cited by* Watson, R.H.J. 1971. Sugar and food choice. *In* Sugar, J. Yudkin, J. Edelman, and L. Hough (Editors). Butterworths, London.

GIFFORD, E. G. 1971. Californian balanophagy. *In* The California Indians, A Source Book, 2nd Edition, R. F. Heizer, and M. A. Whipple (Editors). Univ. of Calif. Press, Berkeley, Calif.

GLANVILLE, E. V., KAPLAN, A. R., and FISCHER, R. 1964. Age, sex and taste sensitivity. J. Geront. *19*, 474-478.

GRACE, J., and RUSSEK, M. 1969. The influence of previous experience on the taste behavior of dogs toward sucrose and saccharin. Physiol. Behav. *4*, 553-558.

HELGREN, F. J. LYNCH, M. J., and KIRCHMEYER, F. J. 1955. A taste panel study of the saccharin "off-taste". J. Am. Pharm. Ass. *44*, 353-355, 442-446.

HELLEKANT, G. 1967. Action and interaction of ethyl alcohol and some other substances on the receptors of the tongue. *In* Olfaction and Taste 2, T. Hayashi (Editor). Pergamon Press, Oxford.

HOCKETT, R. C. 1955. Sugars in human nutrition. *In* Use of Sugars and Other Carbohydrates in the Food Industry. Advances in Chemistry Series, No. 12, Washington, D.C.

HOLT, V. M. 1885. Why Not Eat Insects? E. W. Classey, Ltd., Middlesex.

HUGHES, G. 1969. Changes in taste sensitivity with advancing age. Geront. Clin. *11*, 224-230.

KAPLAN, A. R., GLANVILLE, E. V., and FISCHER, R. 1965. Cumulative effect of age and smoking on taste sensitivity in males and females. J. Geront. *20*, 334-337.

LÁT, J. 1959. *Cited by* Watson, R.H.J. 1971. Sugar and food choice. *In* Sugar, J. Yudkin, J. Edelman, and L. Hough (Editors). Butterworths, London.

LICHTENSTEIN, P. E. 1948. The relative sweetness of sugars: sucrose and dextrose. J. Exptl. Psychol. *38*, 578-586.

LYNCH, G. W. 1971. Social aspects of the taste for sweetness. *In* Sugar, J. Yudkin, J. Edelman, and L. Hough (Editors). Butterworths, London.

MACDONALD, I. 1971. Sucrose and blood lipids. *In* Sugar, J. Yudkin, J. Edelman, and L. Hough (Editors). Butterworths, London.

MARTIN, S., and PANGBORN, R. M. 1970. Taste interaction of ethyl alcohol with sweet, salty, sour and bitter compounds. J. Sci. Fd. Agr. *21*, 653-655.

MAYER-GROSS, W., and WALKER, J. W. 1946. Taste and selection of food in hypoglycaemia. Brit. J. Exp. Pathol. *27*, 297-305.

MCBURNEY, D. H. 1972. Gustatory cross adaptation between sweet-tasting compounds. Percept. and Psychophys. *11*, 225-227.

MEISELMAN, H. L. 1968. Adaptation and cross adaptation of the four gustatory qualities. Percept. and Psychophys. *4*, 368-372.
MEISELMAN, H. L. 1972. Human taste perception. *In* CRC Critical Reviews in Food Technology, T. E. Furia (Editor). *3*, 89-119.
MONCRIEFF, R. W. 1970. The sweet taste. Flavour Ind. *1*, 439-441.
MOORE, M. E., LINKER, E., and PURCELL, M. 1965. Taste-sensitivity after eating: a signal-detection approach. Am. J. Psychol. *78*, 107-111.
MORTON, R. A., and HORNER, A. A. 1961. *Cited by* Bender, A. E. and Damji, K. B. 1971. Some effects of dietary sucrose. *In* Sugar, J. Yudkin, J. Edelman, and L. Hough (Editors). Butterworths, London.
MOSKOWITZ, H. R. 1970A. Ratio scales of sugar sweetness. Percept. and Psychophys. *7*, 315-320.
MOSKOWITZ, H. R. 1970B. Sweetness and intensity of artificial sweeteners. Percept. and Psychophys. *8*, 40-42.
MOSKOWITZ, H. R. 1971. The sweetness and pleasantness of sugars. Am. J. Psychol. *84*, 387-405.
MOSKOWITZ, H. R. 1972A. Perceptual attributes of the taste of sugars. J. Food Sci. *37*, 624-626.
MOSKOWITZ, H. R. 1972B. Perceptual changes in taste mixtures. Percept. and Psychophys. *11*, 257-262.
MOSKOWITZ, H. R. 1972C. Economic applications of sweetness scales. J. Food Sci. *37*, 411-415.
MOSKOWITZ, H. R., and ARABIE, P. 1970. Taste intensity as a function of stimulus concentration and solvent viscosity. J. Texture Studies *1*, 502-510.
NORDSIEK, F. W. 1972. The sweet tooth. Am. Sci. *60* (1) 41-45.
PANGBORN, R. M. 1959. Influence of hunger on sweetness preferences and taste thresholds. Am. J. Clin. Nutr. *7*, 280-287.
PANGBORN, R. M. 1960. Taste interrelationships. Food Res. *25*, 245-256.
PANGBORN, R. M. 1961. Taste interrelationships. II. Suprathreshold solutions of sucrose and citric acid. J. Food Sci. *26*, 648-655.
PANGBORN, R. M. 1962. Taste interrelationships. III. Suprathreshold solutions of sucrose and sodium chloride. J. Food Sci. *27*, 495-500.
PANGBORN, R. M. 1963. Relative taste intensities of selected sugars and organic acids. J. Food Sci. *28*, 726-733.
PANGBORN, R. M. 1967. Some aspects of chemoreception in human nutrition. *In* The Chemical Senses and Nutrition, M. R. Kare and O. Maller (Editors). The John Hopkins Press, Baltimore, Md.
PANGBORN, R. M. 1970. Individual variation in affective responses to taste stimuli. Psychon. Sci. *21*, 125-126.
PANGBORN, R. M. 1971. Relationship of taste and other oral functions to food acceptability. *In* Proceedings, Third International Congress, Food Science and Technology, G. F. Stewart and C. L. Willey (Editors). Institute of Food Technologists, Chicago, Ill.
PANGBORN, R. M., and SIMONE, M. 1958. Body size and sweetness preference. J. Am. Diet. Ass. *34*, 924-928.
PANGBORN, R. M. TRABUE, I. M., and SZCZESNIAK, A. S. 1973. Effect of hydrocolloids on oral viscosity and basic taste intensities. J. Texture Studies *4*, 224-241.
PARISER, E. R. 1961. How physical properties of candy affect taste. Mfg. Confect. *41* (5), 47-50.
PFAFFMANN, C. 1957. Taste mechanisms in preference behavior. Am. J. Clin. Nutr. *5*, 142-147.
PFAFFMANN, C. 1959. The sense of taste. *In* Handbook of Physiology, Vol. 1. 507-534, Am. Physiol. Soc., Washington, D.C.
RENNER, H. D. 1944. The Origin of Food Habits, Faber and Faber, Ltd., London.
RICHTER, C. P., and CAMPBELL, K. H. 1940. Sucrose taste thresholds of rats

and humans. Am. J. Physiol. *128*, 291-297.

SCHIFFMAN, S. S., and ERICKSON, R, P. 1971. A psychophysical model for gustatory quality. Physiol. Behav. 7, 617-633.

SHALLENBERGER, R. S., and ACREE, T. E. 1971. Chemical structure of compounds and their sweet and bitter taste. *In* Taste, L. M. Beidler (Editor). Springer-Verlag, New York.

SHEFFIELD, F. D., and ROBY, T. B. 1950. Reward value of a nonnutritive sweet taste. J. Comp. Physiol. Psychol. *43*, 471-481.

SHIM, K. R. 1968. *Cited by* Watson, R.H.J. 1971. Sugar and food choice. *In* Sugar, J. Yudkin, J. Edelman, and L. Hough (Editors). Butterworths, London.

SJÖSTRÖM, L. B., and CAIRNCROSS, S. E. 1955. Role of sweeteners in food flavor. *In* Use of Sugars and Other Carbohydrates in the Food Industry, Advances in Chemistry Series No. 12, Washington, D.C.

STONE, H., and OLIVER, S. 1966. Effect of viscosity on the detection of relative sweetness intensity of sucrose solutions. J. Food Sci. *31*, 129-134.

U. S. DEPT. OF AGRICULTURE. 1968. Econ. Res. Serv., Food Consumption, Prices, Expenditures. G.P.O., Washington, D.C.

U. S. DEPT. OF AGRICULTURE. 1969. Econ. Res. Serv., National Food Situation, NFS 128, G.P.O., Washington, D.C.

VAISEY, M., BRUNION, R., and COOPER, J. 1969. Some sensory effect of hydrocolloid sols on sweetness. J. Food Sci. *34*, 397-400.

VALENSTEIN, E. S., KAKOLEWSKI, J. W., and COX, V. C. 1967. Sex differences in taste preference for glucose and saccharin solutions. Science *156*, 942-943.

VAN EIJK, A. 1972. The influence of raw materials on the taste of confectionery. Flavour Ind. *3*, 348-352.

VAN HOOK, A. 1949. Sugar, Its Production, Technology, and Uses. Ronald Press, New York.

WATSON, R. H. J. 1964. *Cited by* Watson, R.H.J. 1971. Sugar and food choice. *In* Sugar, J. Yudkin, J. Edelman, and L. Hough (Editors). Butterworths, London.

WATSON, R. H. J. 1971. Sugar and food choice. *In* Sugar, J. Yudkin, J. Edelman, and L. Hough (Editors). Butterworths, London.

YUDKIN, J. 1972. Sweet and Dangerous. Peter H. Wyden, Inc. New York.

YUDKIN, J., EDELMAN, J., and HOUGH, L. 1971. Sugar. Butterworths, London.

CHAPTER 4

Bruce J. Walter | Sweetener Economics

Market Size

The United States sweetener market is a multi-product, multi-industry complex. Among the products generally included in this market are the following: sucrose sweeteners (cane and beet sugar); starch sweeteners (dextrose, conventional corn syrup, high-levulose corn syrup); other caloric sweeteners (honey, maple syrup and sugar, molasses, sugarcane syrup, refiners' syrup); and noncaloric sweeteners (saccharin and others). This list merely indicates the major classifications and types of products included in this market. All these products are available in a wide spectrum of variations, each with very detailed specifications.

While all the products listed above compete with one another as sweeteners, each has a unique set of characteristics—such as flavor, texture, Calorie content, moisture retention, etc.—which differentiates it from the others and makes it especially suitable for particular uses. Thus, the overall sweetener market is, in effect, a collection of overlapping "mini-markets" in which particular sub-sets of these products compete as substitutes or partial substitutes for particular uses.

As indicated in Table 4.1, the U.S. sweetener market has a total value of approximately $2.9 billion at the manufacturer-wholesaler level. In terms of both quantity and value, the sucrose sweeteners (cane and beet sugar) are by far the most important products in this market, accounting for approximately $2.4 billion, or over 82% of its total value. The starch-based sweeteners (corn syrup and dextrose) account for only approximately $0.4 billion (15.5% of the market's total value) and hence are a somewhat distant second in importance. The "other caloric sweeteners" and the noncaloric sweeteners account for only 2.2% and 0.2% of the value of the market, respectively.

Market Trends

The 1961 and 1971 market share and annual per capita consumption of each of the major sweeteners are presented in Table 4.2. These data illustrate the following trends:

(1.) Total per capita sweetener consumption has increased.

TABLE 4.1

THE UNITED STATES SWEETENER MARKET: INDICATED DELIVERIES (BY SOURCE), AVERAGE WHOLESALE PRICE, AND WHOLESALE VALUE, BY TYPE OF SWEETENER, 1971

Type of sweetener	Unit of measure	Indicated domestic deliveries, by source		Total deliveries	Average wholesale price	US Wholesale Value (000 dollars)	Percentage of total value
		Domestically produced	Imported				
Sucrose sweeteners[1]						2,359,708	82.1
Cane sugar (refined)	1,000 cwt	46,685	101,009	147,694	$11.12/cwt	1,642,357	57.2
Beet sugar (refined)	1,000 cwt	64,510	—	64,510	$11.12/cwt	717,351	24.9
Starch sweeteners						444,464	15.5
Corn sugar (dextrose)	1,000 cwt	12,930	—	12,930	$6.86/cwt	88,700	3.1
Corn syrup	1,000 cwt	35,470	—	35,470	$10.03/cwt	355,764	12.4
Other caloric sweeteners						62,240	2.2
Honey	1,000 lb	206,326	[4]3,882	210,208	N.A.	[6]46,917	1.7
Molasses (edible)	1,000 gal	2,517	2,470	4,987	N.A.	N.A.	N.A.
Refiners' syrup	1,000 gal	[2]1,883	[5]—	1,883	N.A.	N.A.	N.A.
Sugarcane syrup	1,000 gal	[2]2,661		[2][5]2,661	N.A.	[2]3,294	0.1
Maple syrup and sugar	1,000 gal	962	577	1,539	N.A.	[6]12,029	0.4

Noncaloric sweeteners							6,827	0.2
Saccharin	1,000 lb	[3]3,000	1,433	4,433	$1.54/lb	6,827	0.2	
Other	—	N.A.	N.A.	N.A.		N.A.	N.A.	
Total							2,873,239	100.0

[1] Assumes that deliveries were in proportion to source of supply as indicated in NFS-143 (i.e., beet 30.4 %, domestic cane 22.0%, and imported cane 47.6%).

[2] 1969 data: series discontinued.

[3] Estimate.

[4] Net imports (i.e., total imports less total exports).

[5] Imported cane syrup included in imported edible molasses.

[6] US farm value plus value of imports.

Sources: Refiners' syrup, maple syrup, honey, and molasses: USDA, *Agricultural Statistics* 1972, pp. 118 and 120.
    Saccharin: USDA-ERS, *National Food Situation*, NFS-143 (February 1973), p. 20.
    Starch sweeteners: USDA-ERS-ESAD, special data (shipments).
    Sucrose sweeteners: USDA-ERS, *National Food Situation*, NFS-143 (February 1973), pp. 20 and 21.
        USDA-ASCS, *Sugar Reports*, No. 238 (March 1972), p. 14.
    Value of imports: Bureau of Census, *1971 Imports* (FT 246-71).

TABLE 4.2

MARKET SHARE AND ANNUAL PER CAPITA CONSUMPTION OF SWEETENERS IN THE UNITED STATES, BY TYPE OF SWEETENER, 1961 AND 1971

Type of sweetener	1961		1971		Difference in market share (1971 minus 1961) (%)
	Per capita consumption (lb)	Market share (%)	Per capita consumption (lb)	Market share (%)	
Sucrose sweeteners	97.7	83.7	102.4	78.0	-5.7
Cane sugar	71.3	61.1	71.7	54.6	-6.5
Beet sugar	26.4	22.6	30.7	23.4	+0.8
Starch sweeteners	14.3	12.3	21.4	16.3	+4.0
Dextrose	3.7	3.2	5.2	4.0	+0.8
Corn syrup	10.6	9.1	16.2	12.3	+3.2
Other caloric sweeteners	2.2	1.9	1.7	1.3	-0.6
Honey	1.3	1.11	1.1	0.84	-0.27
Molasses	0.3	0.26	0.2	0.15	-0.11
Refiners' syrup	0.2	0.17	0.1	0.08	-0.09
Sugarcane syrup	0.2	0.17	0.2	0.15	-0.02
Maple syrup and sugar	0.2	0.17	0.1	0.08	-0.09
Noncaloric sweeteners (sucrose equiv.)[1]	2.5	2.1	5.7	4.4	+2.3
Saccharin (sucrose equiv.)[1]	2.1	1.8	5.7	4.4	+2.6
Cyclamate (sucrose equiv.)[1]	0.4	0.3	0.0	0.0	-0.3
Other	N.A.	N.A.	N.A.	N.A.	N.A.
Total	116.7	100.0	131.2	100.0	0

[1] Assumes saccharin and cyclamate are 300 and 30 times as sweet as sucrose, respectively.

Sources: USDA, ERS, *National Food Situation*, NFS-143 (February 1973), p. 20.
USDA, ERS, *Food Consumption, Prices, and Expenditures*, Agricultural Economic Report No. 138 (July 1968), p. 84.

(2.) The per capita consumption of the sucrose sweeteners has increased, but the market share of these sweeteners has decreased.

(3.) Per capita consumption of both the starch sweeteners and the noncaloric sweeteners has increased, and their relative shares of the market have also increased.

(4.) Both the per capita consumption and the market share of the products classified as "other caloric sweeteners" have declined.

While these observations cover only a recent 11-yr period, the sweetener market has followed these general trends over a much longer time span.

It is interesting to note that while the per capita consumption of both beet and cane sugar increased by 4.7 lb (10.5%) between 1961 and 1971, the increase was much greater for beet sugar (16.3%) than for cane sugar (0.6%). In fact, the market share for beet sugar increased by 0.8 percentage point (from 22.6 to 23.4% of the market), while that of cane sugar decreased by 6.5 percentage points (from 61.1 to 54.6% of the market). Since cane sugar accounts for over 70% of the sucrose sweeteners consumed in this country, the market share of this group of sweeteners also decreased.

While the market share of the sucrose sweeteners decreased by 5.7 percentage points (from 83.7 to 78.0% of the market), that of the starch sweeteners increased by 4.0 percentage points (from 12.3 to 16.3%). Most of this increase was due to a 53% increase in the per capita consumption of corn syrup, which resulted in a 3.2 percentage point increase in the market share of corn syrup. Due to the relatively small per capita consumption of dextrose, a 41% increase in its consumption boosted its market share by only 0.8 percentage point.

Despite the complete ban on the use of cyclamate in food, which was imposed by the Food and Drug Administration (FDA) in late 1969, the per capita consumption and market share of noncaloric sweeteners were both significantly higher in 1971 than in 1961. Per capita consumption of noncaloric sweeteners increased from the equivalent of 2.5 lb of sucrose to 5.7 lb (a 128% increase) between these two years, and the noncaloric segment of the sweetener market increased from 2.1% in 1961 to 4.4% in 1971. The per capita consumption and market share of the noncalorics are, however, both down somewhat from their 1968 peaks of 7.2 lb and approximately 5.6%, respectively.

Industrial Users

One important but often overlooked factor affecting the US sweetener market is the continuing increase in the proportion of

TABLE 4.3

SUGAR DELIVERIES IN HUNDREDWEIGHTS,[1] BY TYPE OF SUGAR AND BY TYPE OF PRODUCT OR BUSINESS OF BUYER, FOR CALENDAR YEAR 1972

Product or business of buyer	Beet	Cane	Imported direct consumption	Total all sugar	Liquid sugar included in totals	
					Beet	Cane
Industrial						
Bakery, cereal and allied products	11,236,889	17,533,797	216,687	28,987,373	298,908	2,039,944
Confectionery and related products	6,979,988	1,402,875	131,201	21,139,064	195,072	3,155,042
Ice cream and dairy products	4,603,347	7,353,141	29,005	11,985,493	2,085,208	5,321,833
Beverages	11,940,071	36,767,372	32,080	48,739,523	6,149,216	23,003,001
Canned, bottled, frozen foods, jams, jellies and preserves	8,845,642	10,806,433	87,062	19,739,137	2,875,535	5,489,181
Multiple and all other food uses	3,712,333	6,332,092	113,402	10,157,827	224,373	1,544,970
Non-food products	359,999	1,430,372	20,701	1,811,072	60,056	654,593
Sub-total	47,678,269	94,251,082	630,138	142,559,489	11,888,368	41,208,564

Non-industrial

Hotels, restaurants, institutions	121,463	1,533,196	38,000	1,692,659	15,618	92,702
Wholesale grocers, jobbers, sugar dealers	12,413,962	29,205,937	436,755	42,056,654	178,506	333,997
Retail grocers, chain stores, super markets	4,792,296	21,237,573	294,111	26,323,980	131,819	189,393
All other deliveries, including deliveries to Government agencies	673,547	1,041,341	42,385	1,757,273	33,217	106,074
Sub-total	18,001,268	53,018,047	811,251	71,830,566	359,160	722,166
Total deliveries	65,679,537	147,269,129	1,441,389	214,390,055	12,247,528	41,930,730
Included in totals:						
Deliveries in consumer-size packages (less than 50 lb)	9,690,019	41,170,595	275,362	51,135,976		
Deliveries in bulk (unpackaged)	30,179,160	35,624,011	3,482	65,806,653		

[1] Reported as produced or imported and delivered except liquid sugar which is on a sugar solids content basis.

Source: USDA-ASCS, *Sugar Reports*, No. 250 (March 1973), p. 21.

total sweeteners distributed to industrial users and the resulting decrease in the proportion distributed for direct home and institutional consumption. This shift in the demand side of the market is extremely important, since it greatly increases the potential for substitution between sweeteners.

Industrial food processors have become the largest users of sweeteners in the United States. In the case of sugar, the major US sweetener and one for which household use has traditionally been a major market outlet, the proportion of deliveries made directly to industrial users has been increasing since at least 1929, the earliest year for which data are available. In that year, only 35.5 million hundredweights, 28% of all the sugar used in this country, was delivered to food processors (USDA 1950). In contrast, over 142.5 million hundredweights of sugar were delivered to industrial users in 1972 (Table 4.3). This represents 66.5% of all sugar deliveries in that year. Only 51 million hundredweights, or 23.9% of all sucrose deliveries, were delivered in consumer-size packages (less than 50 lb) in 1972. Table 4.4 indicates the proportion of sugar delivered to industrial buyers and the proportion delivered in consumer-size packages for each year from 1955 to 1972. It illustrates the

TABLE 4.4

PERCENTAGE OF SUGAR DELIVERIES TO INDUSTRIAL BUYERS AND IN CONSUMER-SIZE PACKAGES, 1955-1972

Year	Deliveries to industrial buyers	Deliveries in consumer-size packages[1]
1955	48.2	NA
1956	49.3	NA
1957	51.2	31.0
1958	49.8	35.3
1959	52.2	34.2
1960	53.6	34.0
1961	53.8	34.2
1962	55.1	31.6
1963	56.1	30.2
1964	57.2	30.7
1965	59.6	29.0
1966	61.4	27.7
1967	62.1	26.9
1968	64.2	25.3
1969	65.1	24.8
1970	65.1	24.0
1971	65.5	24.6
1972	66.5	23.9

[1] Deliveries in packages of less than 50 lb.

Source:　USDA-ASCS, *Sugar Reports*, various issues.

significant and amazingly consistent increase in the proportion of sugar delivered to industrial users and the decrease in the proportion of sucrose delivered in consumer-size packages. Furthermore, it appears that the proportion of the total US sweetener market accounted for by industrial users has increased even more than that indicated by these sugar statistics, since nearly all the dextrose and corn syrup sold in the US are delivered to industrial users (Table 4.5) and the market share of these starch sweeteners has increased significantly.

The increase in the proportion of sweeteners delivered to industrial users and the associated decrease in the proportion delivered directly to households appear to be a reflection of: (1) rising incomes and (2) changes in food habits. A number of studies have shown that direct purchases of sugar per capita are smaller for high-income families than for low-income families

TABLE 4.5

CORN REFINERS' SHIPMENTS OF CORN SYRUP AND DEXTROSE, BY TYPE OF PRODUCT OR BUSINESS OF
BUYER, CALENDAR YEAR 1972 [1]

Type of buyer	Corn syrup, unmixed	Dextrose	All corn sweeteners
	Hundredweights, dry basis[2]		
Bakery, cereal and related products	5,740,326	3,273,525	9,013,851
Confectionery and related products	8,768,286	1,205,053	9,973,339
Ice-cream and dairy products	4,421,012	140,538	4,561,550
Beverages	1,812,724	181,903	1,994,627
Canners and packers, jams, jellies and preserves	3,728,177	633,963	4,362,140
Multiple and all other food uses	6,298,434	3,664,157	9,962,591
Non-food uses[3]	1,227,224	1,803,770	3,030,994
Total shipments [4]	31,996,186	10,902,910	42,899,096

[1] Domestic shipments of members of Corn Refiners Association, Inc.

[2] Assumes a solids content of 80.3% for corn syrup and 92% for dextrose. Thus, dry weight basis is reported weight shipped multiplied by 0.803 for corn syrup and 0.92 for dextrose.

[3] Includes shipments to jobbers (wholesalers).

[4] Due to rounding, totals may not be exact sums of individual items.

Source: USDA-ASCS, *Sugar Reports*, No. 250 (March 1973), p. 23.

(Rockwell 1959). Furthermore, the decline in home baking, canning, and similar activities and the rise in the use of prepared or convenience foods has obviously been an important factor in transferring an increasing proportion of the market for sweeteners from households to industrial food processors.

The importance of the increase in the proportion of sweeteners sold to industrial users is that industrial users are able and willing to substitute one sweetener for another to a much greater extent than household users. The extent of this substitution varies widely among food industries and even among segments of the same industry. For example, the substitution of corn syrup for sugar has been greater in the canning and ice-cream industries than in other industries, and the substitution of noncaloric sweeteners for sugar has been especially important in soft drinks. The limits of substitution between given sweeteners (both in a given product and in aggregate) are determined by the unique technical characteristics of each sweetener and the desired characteristics of the food products in which they are used.

Since sweeteners are a minor item in household expenditures, neither the general level of sweetener prices nor the relative prices of alternative sweeteners is of much interest or consequence to household users. On the other hand, the cost of sweeteners is much more significant to industrial users. General increases in sweetener prices may cause food processors who use significant quantities of sweeteners to make changes in the prices of the products they manufacture. In turn, this may affect their volume of sales and, ultimately, the quantity of sweeteners they purchase. Changes in the relative prices of sweeteners, on the other hand, encourage substitution between sweeteners. Surveys indicate that industrial users are much less concerned with (1) the general level of sweetener prices than with (2) price stability and (3) purchasing sweeteners at prices no higher than those paid by their competitors (Ballinger and Larkin 1964).

In general, the price elasticity of demand for sweeteners in the United States is very low (i.e., consumption is not very responsive to price changes). Of course, since individual sweeteners can be substituted for one another, the elasticity of demand for a given sweetener is somewhat higher than that for all sweeteners as a group. Furthermore, due to their greater willingness and ability to substitute sweeteners, the elasticity of demand for sweeteners delivered to industrial users is somewhat higher than that for sweeteners sold to household users. In other words, for any given sweetener, the quantity demanded by industrial users is more responsive to changes in the relative price of that sweetener than the quantity demanded by household users. Thus, the most important effect of the increase in the proportion of sweeteners sold to industrial users is that it is accompanied by an increase in demand elasticity or responsiveness to price.

Government Regulation

One of the most important exogenous factors affecting the US sweetener market is Government regulation. The variety, quality, quantity, and price of all the products sold in the sweetener market are influenced, either directly or indirectly, by this factor.

The principal areas of Federal legislation affecting the sweetener market are: (1) the Sugar Act, administered by the U.S. Department of Agriculture (USDA), and (2) the Food, Drug, and Cosmetic Act, administered by the FDA. The former effectively controls the quantity and price of the sucrose sweeteners sold in the U.S. market and hence, because of the importance of the sucrose sweeteners, it influences the price and quantity of all sweeteners. The latter controls the quality and types of sweeteners that can be sold in this country and, because of the extensive testing which new products must undergo, acts as a formidable (though necessary) barrier to market entry of new sweeteners.

The Sugar Act.—Actions of the Federal Government have affected the U.S. sugar industry since 1789. A quota system of control was initiated in 1934 and has been amended and extended periodically since then. The current legislation governing the industry is the Sugar Act of 1948 as amended in 1971. This legislation became effective January 1, 1972, and extended the Act through December 31, 1974.

The principal provisions of the United States sugar program are: (1) limitation of the total supply of sugar available to U.S. consumers; (2) Government subsidy payments to U.S. sugarcane and sugarbeet growers; (3) an excise tax on all sugar marketed within the U.S.; and (4) a tariff on sugar imports. The first of these provisions (the limitation of supply) is the major determinant of U.S. sugar prices and, as such, is the only provision to be discussed in this chapter.

Under the supply limitation provision of the Sugar Act, the Secretary of Agriculture each year: (1) determines the quantity of sugar needed to meet the requirements of domestic consumers and to attain the price objective specified in the Sugar Act; (2) divides, by means of quotas, this total supply requirement among specified domestic and foreign production areas; (3) assigns, when necessary for orderly production, "proportionate shares" of each domestic production area quota to individual farms within that production area; and (4) imposes, when necessary for orderly marketing, a refined sugar "marketing allotment" upon each refining and importing firm. Through these strong supply limitation powers granted by the Sugar Act, the Secretary of Agriculture is able to control the price of raw sugar.

The price objective specified in the Sugar Act is to maintain the same ratio between the price of raw sugar, as registered in the New York market, and the average of (1) the parity index (the index of prices paid by all farmers for commodities and services, including interest, taxes, and farm wages, 1967 = 100) and (2) the wholesale price index (1967 = 100) as the ratio that existed during the period September 1, 1970, through August 31, 1971. The Secretary of Agriculture is required to make appropriate adjustments in his determination of national consumption requirements whenever the average price of raw sugar varies from the objective by 4% or more for 7 consecutive days (3% or more during November, December, January, and February).

Under the Sugar Act, domestic sugar prices have generally been significantly higher than those prevailing in the world market (Table 4.6). During the past 25 yr, the world price has exceeded the U.S. price only in 1950-51, 1957, 1963-64, and 1972. It should be noted, however, that although US sugar prices have usually been maintained above the world market level, it does not follow that U.S. prices would have declined to the world level if quotas had not existed and import duties were unchanged. Without quotas, US sugar prices would have declined and those in the world market would have risen until they balanced at some intermediate level.

Although the price of refined sugar is not directly controlled by the Sugar Act, this legislation indirectly influences the price of refined sugar and, in fact, the price of all sweeteners. By controlling the price of raw sugar, the US sugar program establishes a minimum price level (or "floor") for refined beet and cane sugar. Since these two products constitute a very large proportion of the total US sweetener market, their price in turn influences the prices of all other sweeteners.

The generally higher price level for sugar in the U.S. relative to that in the world market is, in effect, the "cost" paid by U.S. sugar consumers for price stability and the protection of the domestic sugar industry. In evaluating the performance of the US sugar program, a recent publication of the University of California (1972), which summarizes the findings of a research project conducted jointly by the USDA and several state agricultural experiment stations, states: "While some consumers have benefited from the

TABLE 4.6

RAW SUGAR PRICES IN NEW YORK AND THE WORLD MARKET, 1948-72, IN CENTS
PER LB

Year	Raw sugar in New York	World Sugar[1]	Differences: New York over world
1948	5.54	5.13	0.41
1949	5.81	5.03	0.78
1950	5.93	5.82	[2]0.11
1951	6.06	6.66	-0.60
1952	6.26	5.08	1.18
1953	6.29	4.27	2.02
1954	6.09	4.14	1.95
1955	5.95	4.19	1.76
1956	6.09	4.47	1.62
1957	6.24	6.10	[2]0.14
1958	6.27	4.36	1.19
1959	6.24	3.86	2.38
1960	6.30	4.09	2.21
1961	6.30	3.85	2.45
1962	6.45	3.87	2.58
1963	8.18	9.41	-1.21
1964	6.90	6.79	[2]0.11
1965	6.75	3.07	3.68
1966	6.99	2.81	4.18
1967	7.28	2.95	4.33
1968	7.52	2.96	4.56
1969	7.75	4.37	3.38
1970	8.07	4.88	3.19
1971	8.52	5.65	2.87
1972	9.09	8.53	[2]0.56

[1] Adjusted to the New York delivery base.

[2] World price exceeded US price in some months.

Source: USDA-ASCS, *Sugar Reports.*

stability of prices, partly offsetting the burden of high prices, other consumers have been placed at some disadvantage by being unable to take advantage of lower seasonal and temporary prices that might have been available with a less regulated market."

In commenting on the effects which the U.S. sugar program has on the prices and market shares of other sweeteners, the publication cited above notes: "For one, producers of corn sweeteners indirectly benefit from high prices for sugar. Such prices for sugar reduce its effectiveness in competing with corn syrup and dextrose." This report could have gone on to say that, in addition to stimulating the substitution of the starch and noncaloric sweeteners already on the market for sucrose, high sugar prices also provide a strong stimulus for the development of new sucrose substitutes.

The Food, Drug, and Cosmetic Act.—As originally enacted, the Food, Drug, and Cosmetic Act of 1938 placed the burden of proving the potential harmfulness of a food additive on the FDA. Under this doctrine, the FDA could take no action against a possibly harmful additive until it was already being used in foods involved in interstate commerce. Then, the FDA had to perform the lengthy testing needed to prove the potential harmfulness and carry its case to the courts in hope of banning the offending substance. As a result of the time involved in this procedure, a food additive might have been in commercial use for months or even years before it could be removed from the food supply.

With the enactment of the Food Additives Amendment on September 6, 1958, the methods of regulating the use of food additives were completely altered. This amendment, which took effect on March 6, 1960, provided that no additive could be used in foods unless the FDA, after a careful review of test data submitted by its manufacturer, agreed that the substance was safe at the intended levels of use. An exception was made for all additives that, because of years of widespread use in foods, were "generally recognized as safe" (GRAS) by experts in the field.

One of the most controversial sections of the entire Food Additives Amendment is that referred to as the Delaney clause. This clause, initiated by Rep. James J. Delaney (D.-N.Y.), prohibits the use of any substance found to cause cancer when ingested by man or animals, regardless of how much of the substance it takes to produce cancer or how long. It is this section of the food and drug laws which required the FDA to ban the use of cyclamate sweeteners in late 1969. Proponents of the cancer clause argue that not enough is known about cancer to establish safe tolerances for carcinogens. Opponents, on the other hand, argue that: (1) even common substances such as glucose and table salt can induce cancer in animals if injected in huge doses and (2) this clause leaves no room for scientific judgment. Despite the controversy which surrounds it, this issue is not likely to be resolved in the near future.

The Food Additives Amendment has had far-reaching effects on the U.S. sweetener industry and sweetener economics. It has: (1) lengthened the time required to develop new sweeteners; (2) increased the cost of developing new sweeteners; and (3) tended to reduce the number of available sweeteners.

At least a 2-yr feeding test using two species of animals (one a rodent and one a nonrodent) is usually required by the FDA under this amendment. In some cases, it may require longer tests and/or the use of additional species of animals. Furthermore, industry sources (Oser 1966) point out: "It's a rule of thumb that a two-year feeding study really takes three years. By the time you run the animal autopsies and perform the tissue examinations at the end of the study, correlate the data, write the reports, submit the evidence to FDA, have conferences with FDA, and work out the myriad of other details, the two years have easily stretched into three."

The cost of the feeding tests and legal work required to obtain FDA approval of a new sweetener under the Food Additives Amendment have significantly increased the cost of developing new products. From the standpoint of economic theory, these additional costs act as formidable (though necessary) barriers to market entry.

The types and variety of available sweeteners and, hence, the level of competition in the sweetener market, have been affected by the Food Additives Amendment both: (1) directly, through bans or limits imposed on existing sweeteners, and (2) indirectly, through the additional time and cost required to develop new products. In late 1969, under the provisions of the Delaney clause, the FDA imposed a complete ban on the use of cyclamate sweeteners in food. In 1972 saccharin was removed from the GRAS list and limits were placed on the amounts of saccharin permitted in foods and beverages. The latter action does not prohibit food processors or bottlers from using saccharin, but it is intended to "freeze" its use at current levels until it can be further tested.

While the direct effects of the Food Additives Amendment on the U.S. sweetener market can be readily cited and have certainly been significant, the indirect effects, although much more difficult to grasp or measure, may be even more significant and far-reaching. For example, no one will ever know: (1) just how much research and development effort has been redirected from sweeteners and other food additives to completely different types of products or (2) the number and potential value of sweeteners discovered but not developed because of the time and cost barriers imposed on market entry by the Food Additives Amendment. This is not to suggest that the amendment has stifled research completely. On the contrary, sweetener research appears to be moving ahead at a reasonably rapid pace. However, the emphasis or direction of this research appears to have shifted toward the investigation of compounds that are naturally present in foods or are related to such compounds. The reason for this shift is that the FDA's testing requirements for such compounds are, understandably, less stringent and hence less costly than those for totally new synthetic compounds. The barriers imposed by the Food Additives Amendment also seem to have discouraged development of sweeteners that are discovered but found to be not significantly better than products already on the market. It is simply unprofitable to spend large amounts of money to safety-test a sweetener if it is just a "me-too" item that, at best, may capture only a very small share of the total market. Thus, the amendment has created the need for greater selectivity in product development and, as a result, has reduced the number of potential competitors in the market.

The Future

Consumer diet consciousness and the desire of industrial food processors for lower-cost sweeteners have stimulated the search for sugar substitutes. The ban on cyclamates and the limitations placed on the use of saccharin have probably intensified this search. Although a large number of sweetener "discoveries" have been announced in recent years, most of them are laboratory curiosities of relatively little commercial significance.

Among the many new sweeteners recently investigated are a number of substances derived from tropical fruits, such as the miracle fruit, the serendipity berry, and katemfe (Inglett 1971). Others have been derived from such diverse sources as pine tree rosin (Tahara et al. 1971) and citrus peel (Horowitz and Gentili 1969). One thing that all these sweeteners have in common, however, is that they have not to date had any impact on the sweetener market. In fact, none of them has yet completed the toxicity tests required by the FDA for entry into the US sweetener market. Furthermore, a number of these exotic and widely publicized sweeteners have characteristics, such as persistent aftertastes, which make them undesirable for most commercial uses.

One recent discovery, aspartame (aspartyl-phenylalanine methyl ester), is nearing completion of the required tests and will probably be marketed in the not-too-distant future. However, since this new noncaloric sweetener is subject to instability and loss of sweetening power when stored in aqueous solutions, it will be marketed only for tabletop use and for use in dry foods, such as cereals, powdered drink mixes, gelatin, and frosting preparations. Thus, due to the limitations in the ways in which aspartame can be used, it does not appear that there will be another noncaloric sweetener available to take the place of saccharin if saccharin is banned within the next year or two. On the other hand, in light of the current research effort in the field, the situation appears likely to change considerably in the somewhat longer run.

It should be noted that, should saccharin be banned before a suitable replacement is developed and tested, the short-run impact on the sweetener market would be considerably greater than that which followed the cyclamate ban. In the first place, cyclamate users had saccharin to fall back on. Furthermore, since its introduction in this country more than 80 yr ago, saccharin has been the major noncaloric sweetener on a sucrose equivalence basis. Even during the peak cyclamate consumption years of 1968 and 1969, saccharin provided over 75% of the noncaloric sweetening power consumed in this country. On the other hand, it should also be noted that the presence of saccharin, which is extremely low in cost on a sweetness equivalence basis, may in fact have been a major deterrent to the development of other, more costly noncaloric sweeteners (Walter 1972).

In the caloric sweetener field, the economic outlook is considerably different. Momentum has increased in the development and marketing of high-levulose corn syrups. These new syrups, produced by the partial isomerization of dextrose into levulose (also known as fructose), are said to be comparable in their sweetening properties to sucrose. Thus, unlike previous corn syrups, they may be used as complete substitutes for liquid sucrose. This means that these new syrups have access to an extremely important segment of the sweetener market. In 1972, approximately 25% of all sugar deliveries were made in liquid form (Table 4.3). In addition, a significant (but unknown) amount of this sugar was delivered dry, but was liquefied prior to use; hence, high-levulose syrup probably could have been used in its place.

Nationwide, the cost of high-levulose syrups is reportedly about 10% below that of sugar. Industry sources estimate that current production of these new starch-based syrups is equivalent to approximately 3.5 million hundredweights of sugar (dry basis) per yr (Rosenbaum 1972). This is a significant quantity for such a relatively new product. It is equal to approximately 1.6% of the total sugar deliveries or 6.5% of the total liquid sugar deliveries in 1972 (Table 4.3). In light of their marketing performance to date, it appears that high-levulose syrups will have a significant impact on the US sweetener market in the not-too-distant future.

BIBLIOGRAPHY

BALLINGER, R. A. 1971. A history of sugar marketing. AER 197. USDA, ERS. Washington, D.C.
BALLINGER, R. A. 1971. The structure of the U.S. sweetener industry. AER 213. USDA, ERS. Washington, D.C.
BALLINGER, R. A. 1969. Sugar substitutes. *In* Synthetics and substitutes for agricultural products: a compendium. Misc. Pub. 1141. USDA, ERS. Washington, D.C.
BALLINGER, R. A., and LARKIN, L. C. 1964. Sweeteners used by food processing industries: their competitive position in the United States. AER 48. USDA, ERS. Washington, D.C.
BUREAU OF CENSUS. 1972. 1971 imports. FT 246-71. Washington, D.C.

HOROWITZ, R. M., and GENTILI, B. 1969. Taste and structure in phenolic glycosides. J. Agr. Food Chem. *17*, 696-700.

INGLETT, G. E. 1971. Intense sweetness of natural origin. *In* Sweetness and Sweeteners. G. G. Birch, L. F. Green, and C. B. Coulson (Editors). Applied Science Publishers Ltd., London.

OSER, B.L. 1966. *Cited by* H. J. Sanders. Food Additives. Chem. Eng. News 44: Oct. 10, 100-120 and Oct. 17, 108-128.

ROCKWELL, G. R. 1959. Income and household size: their effects on food consumption. MRR 340. USDA, ERS. Washington, D.C.

ROSENBAUM, C. 1972. Corn sweetener termed technical breakthrough. J. Comm., Nov. 8, 6.

TAHARA, A., NAKATA, T., and OHTSUKA, Y. 1971. New type of compound with strong sweetness. Nature *233*, 619-620.

UNIV. OF CALIF. 1972. Economic behavior in the United States sugar market. Calif. Ag. Expt. Sta. Bull. 859.

USDA, 1973. Agricultural statistics 1972. Washington, D.C.

USDA, ASCS. 1973 and earlier. Sugar reports. Washington, D.C.

USDA, BAE. 1950. The world sugar situation. Washington, D.C. Dec., p. 54.

USDA, ERS. 1968. Food, consumption, prices, and expenditures. AER 138. Washington, D.C.

USDA, BAE. 1950. The world sugar situation. Washington, D.C. Dec., p. 54.

WALTER, B. J. 1972. Noncaloric sweeteners. Natl. Agr. Outlook Conf., Washington, D.C., Feb. 24.

CHAPTER 5

A. J. Vlitos | Sucrose

It is timely to be writing about sucrose when there is considerable concern about shortages of conventional sources of energy. For sucrose represents a unique substance—one of the major products resulting from the conversion of solar to chemical energy by higher plants. This fact is bound to be of increasing significance, when comparing sucrose with other sweeteners, since most nonbiological, synthetic sweeteners require an input of energy in their synthesis, and in the long run may compete for supplies of oil, coal and electricity—unless of course someone learns how to synthesize a new sweetener employing solar energy! Thus, irrespective of the discovery of new, synthetic sweeteners, sucrose will continue to be a leading commodity in world trade, for at least three reasons: (1) its relatively efficient biosynthesis by sugarcane and sugar beet, (2) its special chemical and physical properties, and (3) its wide acceptance as a food, and versatility as an additive to other foods.

In view of its rather diverse characteristics, it is difficult to review any single facet of sucrose in depth. Therefore this chapter will attempt to deal with those themes which are likely to become increasingly significant, namely the biosynthesis of sucrose and the properties which make it attractive as a food.

BIOLOGY OF SUCROSE

General

Green plants are characterized by their ability to convert carbon dioxide and water to carbohydrate by photosynthesis. The carbohydrate thus formed provides the energy required for most of the vital processes in the plant—from translocating its own energy throughout its system to the uptake of water and minerals from the soil, as well as for the synthesis of proteins, nucleic acids, hormones, and enzymes. Some of the carbohydrate formed as a result of photosynthesis provides structural materials such as cellulose and lignin. The carbohydrate remaining after the vital processes have been satisfied is stored. Most plants store starch, sucrose or inulin (Edelman 1960).

Sucrose occupies a key position in the metabolism of green plants because it is not only an early product of photosynthesis, but it is also the main form in which carbon is moved about in the plant. The sugar functions as a relatively stable, nonreactive osmotic agent and helps to maintain the turgor or rigidity of plant tissues. Its presence in plant cells has been suggested as a means of protecting the cells from frost damage by stabilizing proteins against denaturation (Andrews and Levitt 1967).

Although sucrose plays a key role in the metabolism of green plants, it is relatively less important in nongreen plants and in animals. Movement of carbohydrates in animals occurs mainly as glucose, as for example in the circulatory system of man and other mammals. In insects and in several other invertebrates the reserve carbohydrate is another nonreducing disaccharide, trehalose (a, a' -diglucoside). Trehalose has also been reported as a storage sugar in fungi, algae and in pteridophytes (Wyatt 1967). The alimentary enzyme which hydrolyzes sucrose in animals is an α-glucosidase (or glucoinvertase) while in plants (yeasts) it is a β-fructosidase (Gilmour 1961).

Biosynthesis

The major commercial production of sucrose depends upon the cultivation of plants such as sugarcane and sugar beet, which store sucrose as a reserve. Both of these species are characterized by a convenient storage organ (or "sink") in which the sucrose is accumulated. In sugarcane the relatively large stalk, and in sugar beet a well-developed storage root act as "sinks". Plant breeders have selected new varieties on the basis of sugar content, as well as on the relative size of the storage organs.

Although it has been thought for many years that sucrose is synthesized in the chloroplasts (Everson, Cockburn, and Gibbs 1967), new evidence suggests that the sugar is in fact formed in the cytoplasm surrounding the chloroplasts (Walker 1970). Certain metabolites appear to be able to move in and out of the chloroplasts and the rate of photosynthesis of isolated chloroplasts is governed by a balance between ribose-5-phosphate and orthophosphate, both of which in solution flow freely into the chloroplast.

The classic photosynthetic carbon reduction cycle (or "Calvin cycle") (Bassham and Calvin 1957) involves the carboxylation of 3 ribulose diphosphate molecules to produce 6 triose phosphate molecules; 5 of which continue the cycle, while the sixth triose phosphate molecule is channelled toward the biosynthesis of secondary products (i.e. carbohydrates, fats, amino acids). In sugarcane and in certain other tropical grasses with dimorphic chloroplasts the C_4-dicarboxylic acid photosynthetic pathway has been reported (Hatch and Slack 1968; Laetsch 1969), but the end result is the same as in the Calvin cycle (Fig. 5.1).

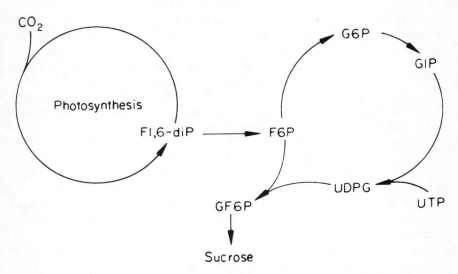

FIG. 5.1. THE BIOSYNTHETIC PATHWAY TO SUCROSE IN THE LEAF

In the synthesis of carbohydrate by higher plants, 2 triose phosphates condense to form fructose-1, 6-diphosphate (FDP), which then loses phosphate to become fructose-6-phosphate (F6P). During the synthesis of sucrose, the nucleotide sugar uridine diphosphate glucose (UDPG), itself derived from F6P and the nucleotide uridine triphosphate (UTP), acts as a glucosylating agent. Thus a glucose residue is transferred to F6P. This transfer is catalyzed by an enzyme, sucrose phosphate synthetase (Hassid 1967). The resulting sucrose monophosphate is broken down by sucrose phosphatase to give free sucrose and phosphate (Hatch 1964; Haq and Hassid 1965).

It is known also that sucrose may be synthesized in some plants by reacting UDPG and fructose to yield sucrose and UDP, i.e.

$$UDPG + fructose \rightleftharpoons sucrose + UDP$$

This is a freely reversible reaction catalyzed by sucrose synthetase, and could account for degradation of sucrose in certain cases (Milner and Avigad 1965).

Translocation and Movement

It is not generally recognized that sucrose is *the* main mobile energy source in higher plants. All carbohydrates synthesized in the leaves are converted to sucrose before they are moved to other parts of the plant (Swanson 1959). In a few cases more complex sugars, such as stachyose, verbascose and raffinose, are also moved about in plants, but to date free hexoses do not appear to be translocated by higher plants (Zimmerman 1957).

The concentration of sucrose in plant sap is rather high, ranging from 10% to 25% (w/v) in different species and at different times of the day. Since sucrose has to move from the leaves to the stems and roots through the sieve tubes of the phloem — a series of elongated cells connected at their ends by perforations of the cell walls known as sieve plates — it is subject to metabolic attack. Sucrose is less reactive than the hexoses and is therefore more stable in a cytoplasmic environment. It has been postulated (Arnold 1968) that its unique role in the translocation of carbohydrate is as a "protected derivative" of glucose. In this respect, sucrose possesses advantages over maltose and turanose, as well as over the disaccharide, trehalose, which is hydrolyzed more easily.

The free energy of hydrolysis of sucrose is -6.6 kcal/mole. This is higher than that of maltose or lactose (-3.0 kcal/mole) and, in fact, approaches those of "high-energy molecules," such as adenosine triphosphate (-6.9 kcal/mole) and UDPG (-7.6 kcal/mole).

As might be expected, since sucrose is a major translocatable compound, enzymes capable of splitting the sugar occur in most plants. Different types of invertases are found in mature and immature sugarcane stems. These enzymes exert control over accumulation or utilization of sucrose for further growth (Hatch and Glasziou 1963; Sacher, Hatch, and Glasziou 1963; Hawker and Hatch 1965). Invertase synthesis has been shown to be under the regulation of hormonal activity (Glasziou, Waldron, and Bull 1966; Gayler and Glasziou 1969). Thus environmental factors may influence the growth of sugarcane indirectly through their effects upon hormone synthesis and the movement of plant hormones from one portion of the plant to another.

Accumulation

One of the most important practical questions facing sugarcane and sugar beet growers is how to ensure that the sucrose accumulated in cane or beet may be prevented from being lost or from undergoing changes before the crop is ready for harvest.

Sucrose formed during photosynthesis is either translocated to the growing regions of the plant where it acts as a substrate for respiration and structural growth, or it moves to mature storage tissues where it is transformed to reserve polysaccharides (such as starch or inulin) or is stored directly as sucrose. One may say that the organic substances of the plant are derived, directly or indirectly, from sucrose.

In sugarcane the process of storing reserve carbohydrate (as sucrose, starch, etc.) and the process of growth appear to be inversely related, because they both draw upon the sugars produced during photosynthesis. These processes are regulated by interactions between genetic and environmental factors which, in turn, determine the rate of photosynthesis and the rate of growth, and thus the rate of accumulation of sucrose. In recent years, certain types of growth-regulating chemicals have been reported to modify the genetic-environmental interactions, and such "chemical ripening" agents may be feasible for commercial use in sugarcane agriculture (Fewkes 1970; Yates and Bates 1958; Glasziou 1964; Nickell and Tanimoto 1965; Vlitos and Lawrie 1967; Alexander 1969).

SOME CHARACTERISTICS OF SUCROSE AS A FOOD

Metabolic Factors

Carbohydrates, in general, are known to be essential in animal metabolism for the assimilation of amino acids and proteins (Cuthbertson and Munro 1939), which otherwise are deaminated in the liver to produce glucose and lipids. The average daily protein requirement of an adult is about 1 gm/kg and the energy expenditure between 1700 and 5000 Calories, according to environment, activity and basal metabolic rate; 1 lb of refined sucrose provides 1800 Calories and represents the least expensive source of assimilable energy currently available (Sugar Year Book 1968).

In animals, sucrose is more rapidly absorbed than starch, since starch must be digested by a two-stage, enzymatic hydrolysis involving ptyalin in the saliva and amylase in the duodenum. Sucrose passes straight through the stomach to the small intestine where hydrolysis to glucose and fructose, and assimilation take place concurrently (Gray and Ingelfinger 1966; Widdas 1970). Glucose is absorbed very rapidly. It can pass into the bloodstream against a concentration gradient and about twice as fast as fructose (Fordtran and Ingelfinger 1968). It is generally thought that fructose is metabolized in the liver, where it is converted to glucose, glycogen and triglycerides; but there may be other metabolic paths involved (Macdonald).

Glucose is metabolized in extra-hepatic tissues, giving rise to direct deposition of fat. Ingested fructose, unlike glucose, leads to an enhanced plasma triglyceride concentration (hyperlipidaemia) (Naismith 1970). Serum cholesterol levels have been reported to be increased following ingestion of sucrose (Winitz, Graff, and Seedman 1964; Lopez, Hodges, and Krehl 1966). These observations, often based on abnormally high levels of sucrose, have led to the proposition by Yudkin (Yudkin 1967A, 1967B, 1969; Yudkin and Roddy 1964) that sucrose is detrimental to health and is a factor in the incidence of ischaemic heart disease. However, Truswell has indicated that when 80 to 85% of the diet is composed of sucrose (an overwhelmingly imbalanced diet!) and 5% of the diet is composed of fat, there is a moderate increase in plasma triglycerides in normolipaemic subjects (in males only). When only 25% of the diet is composed of sucrose and 40% to 45% of fat, normolipaemic subjects do not exhibit any change in plasma triglyceride levels. Undoubtedly there is good sense in the old ethic "eat a little of everything and not too much of any one thing".

In terms of calorific values, sucrose has a lower calorie equivalent than pure starch, protein, or fat *on a dry-weight basis.* However, on a fresh-weight basis, sucrose is approximately equivalent to fat, mutton, or pork in terms of calories.

One important factor favoring the incorporation of sucrose in the diet is that it avoids introducing nutritional imbalances, such as may result if saturated animal fats are to be substituted for the equivalent quantity of sucrose. A natural demand for sugar has been indicated by preferences for sweet foods by subjects under stress (Stare 1960; Watson 1970), as for example in athletes, cosmonauts, etc. Sucrose fed to animals prior to transport and slaughter counters the deterioration in meat quality induced by stress.

Consumption

Sucrose consumption has been suggested (Yudkin 1964A, 1964B) to be an index of affluence. A more comprehensive comparison of the annual per capita consumption of sucrose (Sugar Year Book 1968) does not support this proposition. Cuba, The Netherlands and the Republic of Ireland show the highest consumption (Fig. 5.2)—approximately three times that of Japan. Obviously, dietary habits are strongly influenced by historical, social and commercial factors, while economic factors allow preferences for certain types of food to be exercised. Thus, an affluent society will tend to choose the more expensive animal-derived foods, such as meat, butter, cream, and fish. Although in most of the developing nations, sucrose consumption is relatively low, this situation is changing rapidly, as many of the tropical nations are establishing local sugar industries (Fig. 5.3).

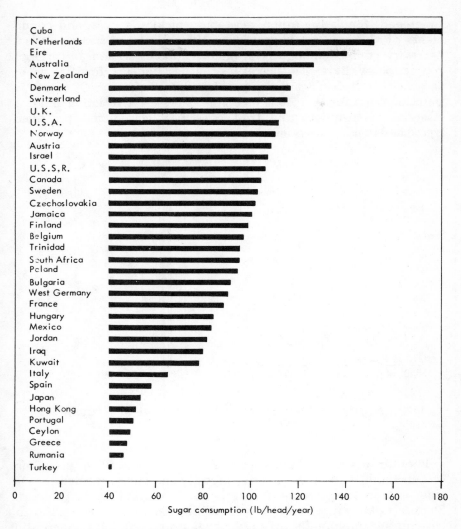

FIG. 5.2. ANNUAL NATIONAL CONSUMPTION OF SUGAR PER HEAD

Food Applications

Apart from its value as a source of energy, sucrose has properties that make possible a rather wide variety of forms in which foods are available. Since it is the second sweetest sugar after β-fructose (Shallenberger 1971), and has a sweet taste with no secondary or after-flavors (Brook 1970), it makes many foods palatable which otherwise might be bland, sour, or unpleasant in other ways. Thus, citrus juices enjoy greater popularity as drinks when they are sweetened. Similarly, fruits, vegetables, pickles, sauces, and condiments of all types are improved by the addition of sucrose. It is a necessary component of cakes, biscuits, puddings, and all types of confectionery. Beverages such as beer, wines, liqueurs, fortified wines and many soft drinks are also sweetened with sucrose. Of course, a traditional use for sucrose is based on its preservative action and it is essential in manufacturing fruit conserves, glazed and preserved fruits, and in condensed milk.

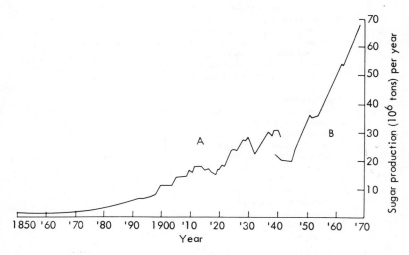

FIG. 5.3. TOTAL WORLD SUGAR PRODUCTION A: ALL SUGAR; B: CENTRIFUGAL SUGAR

Because of its physical properties, sucrose is useful in providing certain foods with texture, body, viscosity and moisture retention. It aids in preventing syneresis in gels and denaturation of proteins. Sucrose is known to assist the emulsification of fats, as well as to develop and modify flavors either by autolysis or by a synergistic action with salt (Pangborn 1962) or citric acid (Pangborn 1960, 1963).

In the U.S., the annual per capita consumption of sucrose (about 100 lb) exceeds the intake of sugars from all other sources (19.1 lb) (U.S. Dept. of Agriculture). The ratio (89.4%) is similar to that for Great Britain (87.5%) (International Sugar Council 1963). Obviously, sucrose is clearly well established as an important food that will not be easily replaced. Sucrose will have to be considered in future as a relatively inexpensive and inexhaustible source of energy now that the more conventional energy sources are under stress (Fig. 5.4).

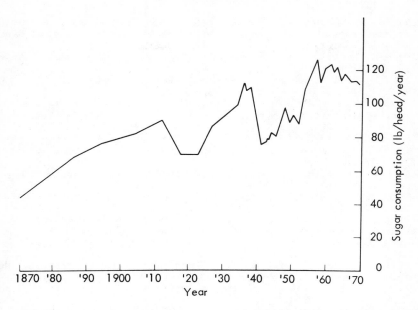

FIG. 5.4. ANNUAL CONSUMPTION OF SUGAR IN GREAT BRITAIN OVER
THE LAST CENTURY

BIBLIOGRAPHY

ALEXANDER, A. G. 1969. The use of chemicals for sucrose control in sugar
cane. Sugar y Azucar *64*,(2) 21.

ANDREWS, S. and LEVITT, J. 1967. Effect of cryoprotective agents on
intermolecular SS formation during freezing of Thiogel. Cryobiology *4*, 85.

ARNOLD, W. N. 1968. The selection of sucrose as the translocate of higher
plants. J. Theor. Biol. *21*, 13.

BASSHAM, J. A. and CALVIN, M. 1957. The path of carbon in photosynthesis.
Prentice-Hall, New York.

BOARD OF TRADE JOURNAL. 1968. *195*, 40.

BROOK, M. 1970. Sucrose and the food manufacturer, p. 32. *In:* Sugar J.
Yudkin, J. Edelman and L. Hough, Editors. Butterworths, London.

CUTHBERTSON, D. P. and MUNRO, H. N. 1939. XV. The Relationship of
carbohydrate metabolism to protein metabolism. I. The roles of total dietary
carbohydrate and of surfeit carbohydrate in protein metabolism. Biochem. J.
33, 128.

EDELMAN, J. 1960. Transfructosylation in plants and especially in *Helianthus
tuberosus* L. Bull. Soc. Chim. Biol. *42*, 1737.

EVERSON, R. G., COCKBURN, W., and GIBBS, M. 1967. Sucrose as a product
of photosynthesis in isolated spinach chloroplasts. Plant Physiol. *42*, 840.

FEWKES, D. W. 1970. Plant growth regulators and the accumulation of sucrose
in sugar cane. *In* Sugar p. 133. J. Yudkin, J. Edelman, and L. Hough, Editors.
Butterworths, London.

FORDTRAN, J. C. and INGELFINGER, F. J. 1968. Absorption of water,
electrolytes and sugars from the human gut. Handbook of Physiology, Section
6, Vol. 3, p. 1475. American Physiological Society, Washington, D.C.

GAYLER, K. R., and GLASZIOU, K. T. 1969. Plant enzyme synthesis: hormonal regulation of invertase and peroxidase synthesis in sugar cane. Planta *84*, 185.

GILMOUR, D. 1961. The Biochemistry of Insects, p. 42. Academic Press, London.

GLASZIOU, K. T. 1964. Rep. David North Plant Res. Centre (1964), p. 48. Colonial Sugar Refining Company, Australia.

GLASZIOU, K. T., WALDRON, J. C., and BULL, T. A. 1966. Control of invertase synthesis in sugar cane. Loci of auxins and glucose effects. Plant Physiol. *41*, 282.

GRAY, G. M., and INGELFINGER, F. J. 1966. Intestinal absorption of sucrose in man: Interrelation of hydrolysis and monosaccharide product absorption. J. Clin. Invest. *45*, 38.

HAQ, S. and HASSID, W. Z. 1965. Biosynthesis of sucrose phosphate with sugar cane leaf chloroplasts. Plant Physiol. *40*, 591.

HASSID, W. Z. 1967. Transformation of sugars in plants. Ann. Rev. Plant Physiol. *18*. 253.

HATCH, M. D. and GLASZIOU, K. T. 1963. Sugar accumulation cycle in sugar cane. II. Relationship of invertase activity to sugar content and growth rate in storage tissue of plants grown in controlled environments. Plant Physiol. *38*, 344.

HATCH, M. D. 1964. Sugar accumulation by sugar-cane storage tissue: the role of sucrose phosphate. Biochem. J. *93*, 521.

HATCH, M. D. and SLACK, C. R. 1968. A new enzyme for the interconversion of pyruvate and phosphopyruvate and its role in the C_4-dicarboxylic acid pathway of photosynthesis. Biochem. J. *106*, 141.

HAWKER, J. S., and HATCH, M. D. 1965. Mechanism of sugar storage by mature stem tissue of sugarcane. Physiologia Plant *18*, 444.

INTERNATIONAL SUGAR COUNCIL. 1963. Table—Production of centrifugal sugar. The World Sugar Economy Structure and Policies, Vol. 2, p. 230.

LAETSCH, W. M. 1969. Relationship between chloroplast structure and photosynthetic carbon-fixation pathways. Sci. Prog., Oxf. *57*, 323.

LOPEZ, A., HODGES, R. E., and KREHL, W. A. 1966. Some interesting relationships between dietary carbohydrate and serum cholesterol. Am. J. Clin. Nutr. *18*, 149.

MACDONALD, I. (personal communication)

MILNER, Y., and AVIGAD, G. 1965. Thymidine diphosphate nucleotides as substrates in the sucrose synthetase reaction. Nature, London *206*, 825.

NAISMITH, D. J. 1970. The hyperlipidaemic property of sucrose. *In Sugar*, p. 183. J. Yudkin, J. Edelman and L. Hough, Editors. Butterworths, London.

NICKELL, L. G., and TANIMOTO, T. 1965. Effects of chemicals on ripening of sugarcane. Rep. Ann. Conf. Hawaiian Sugar Technol. *24*, 152.

PANGBORN, R. M. 1960. Taste interrelationships. Food Res. *25*, 245.

PANGBORN, R. M. 1962. Taste interrelationship. III. Suprathreshold solutions of sucrose and sodium chloride. J. Food Sci. *27*, 495.

PANGBORN, R. M. 1963. Relative taste intensities of selected sugars and organic acids. J. Food Sci. *28*, 726.

SACHER, J. A., HATCH, M. D., and GLASZIOU, K. T. 1963. Sugar accumulation cycle in sugar cane. III. Physical and metabolic aspects of cycle in immature storage tissues. Plant Physiol. *38*, 348.

SHALLENBERGER, R. S. 1971. The theory of sweetness. *In*: Sweetness and Sweeteners, p. 43. G. G. Birch, L. F. Green and C. B. Coulson, Editors. Applied Science, London.

STARE, F. J. 1960. Diets for athletes. J. Am. Dietet. Ass. *37*, 371.

SUGAR YEAR BOOK. 1968. Table—Retail prices of white refined sugar in selected countries. International Sugar Organization, p. 393.

SWANSON, C. A. 1959. Translocation of organic solutes. *In*: Plant Physiology, a Treatise. F. C. Steward, Editor. Vol. 2, p. 481. Academic Press, New York.

TRUSWELL, A. S. 1972. Human nutritional problems at four stages of technical development. Queen Elizabeth College Magazine No. 4. pp. 1-9 (Reprinted from Inaugural Lecture, May 1972).

U.S. DEPARTMENT OF AGRICULTURE. 1970. Table—U.S. Sweeteners: selected supply and distribution data 1967-69. National Food Situation, p. 19.

VLITOS, A. J., and LAWRIE, I. D. 1967. Chemical ripening of sugarcane. A review of field studies carried out in Trinidad over a five year period. Proc. Int. Soc. Sug. Cane Tech. *12*, (1965) 429.

WALKER, D.,A. 1970. The site of sucrose synthesis in green plants. *In: Sugar* p. 103. J. Yudkin, J. Edelman and L. Hough, Editors. Butterworths, London.

WATSON, R. H. J. 1970. Sugar and food choice. *In:* Sugar p. 24, J. Yudkin, J. Edelman and L. Hough, Editors. Butterworths, London.

WIDDAS, W. F. 1970. The role of the intestine in sucrose absorption. *In*: Sugar p. 155. (J. Yudkin, J. Edelman and L. Hough, Editors). Butterworths, London. 1970.

WINITZ, M., GRAFF, J., and SEEDMAN, D. A. 1964. Effect of dietary carbohydrate on serum cholesterol levels. Archs. Biochem. Biophys. *108*, 576.

WYATT, G. R. 1967. The biochemistry of sugars and polysaccharides in insects. Adv. Ins. Physiol. *4*, 287.

YATES, R. A., and BATES, J. F. 1958. Preliminary experiments on the effects of chemicals on the ripening of sugarcane. Proc. Brit. W. Indies Sug. Technol. (1957), 174.

YUDKIN, J. 1964A. Dietary fat and dietary sugar in relation to ischaemic heart-disease and diabetes. Lancet ii, 4.

YUDKIN, J. 1964B. Patterns and trends in carbohydrate consumption and their relation to diseases. Proc. Nutr. Soc. *23*, 149.

YUDKIN, J. 1967A. Sugar and coronary thrombosis. New Scientist *33*, 542.

YUDKIN, J. 1967B. Why blame sugar? Chem. & Ind. 1464.

YUDKIN, J. 1969. Sucrose: heart disease culprit. Chem. Eng. News *47*,(26) 18.

YUDKIN, J., and RODDY, J. 1964. Levels of dietary sucrose in patients with occlusive atherosclerotic disease. Lancet ii, 6.

ZIMMERMAN, M. H. 1957. Translocation of organic substances in trees. I. The nature of the sugars in the sieve tube exudate of trees. Plant Physiol. *32*, 288.

CHAPTER 6

P. K. Chang

Sucrose Chemicals and Their Industrial Uses

Sucrose, even in highly industrialized countries, is still over-whelmingly used as food. Approximately 98% is consumed as a sweetener of processed or directly consumed foods, while less than 2% of the total sugar deliveries in the United States are directed to nonfood uses (Anon. 1971A). An expansion of sucrose utilization beyond food areas can only be sought in new and broad chemical and industrial applications.

From our understanding of sucrose chemistry, the molecule of this disaccharide contains 3 primary and 5 secondary hydroxyl groups. It is possible to replace one or more of these groups by other substituents to obtain a variety of derivatives. In the simplest case, if only one hydroxyl group is replaced by one substituent, theoretically 255 different compounds could result. In the most complex case, all 8 hydroxyl groups can be replaced by 8 different kinds of substituents, resulting in 16,000,000 different compounds. In both cases, only a few major reaction products predominate, while other possible products may exist only in insignificant amounts. Thus, so far, in actual practice, no more than 100 well-defined sucrose derivatives can be prepared. The complexity of the chemical reaction of sucrose lies in the relative reactivity of the 8 displaceable hydroxyl groups. In general, the primary hydroxyl groups are much more reactive than the secondary hydroxyl groups. Yet, a selective substitution restricted to these primary groups was reported successful with a special treatment only 10 yr ago (Barker *et al.* 1963).

There are several interesting characteristics of sucrose derivatives. The polyhydroxyl groups impart water solubility or hydrophilic properties to sucrose derivatives. The polyhydroxyl groups of sucrose derivatives, after being replaced by various functional substituents, when incorporated into sucropolymers, possess "polyfunctionality". Compounds containing high sucrose units are susceptible to attack by microorganisms and are decomposed similarly by a fermentation process. This property, known as. "biodegradability", has certain advantages in the practical use of sucropolymers.

The failure to obtain desired products in many sucrose reactions is due to the instability of the glycosidic linkage, which causes inversion to take place before the intended reaction starts. Failure may also be due to steric hindrance imposed on some of the hydroxyl groups, making complete substitution difficult. The low thermostability of sucrose derivatives or monomers containing sucrose units may also account for many unsuccessful polymerization reactions if high temperature is required. Since several isomers or homologues are produced in one chemical reaction, the isolation and purification of a specific compound can often be tedious and time-consuming. Only the octa-substituted or some partially substituted derivatives are chemically well-defined.

The International Sugar Research Foundation (ISRF) has made great efforts to promote the study of fundamental sucrose chemistry by supporting research projects at the University of London, led by Professor Leslie Hough, in chemical studies: (1) preparation of new sucrose derivatives (Hough 1971); (2) techniques for isolating products; (3) proof of their structures; and (4) reaction mechanisms. Successful methods have been established for the synthesis of mono-, di-, tri-, and tetra-functional derivatives of sucrose, including deoxy, halogeno, thio and amino derivatives. The gain in fundamental scientific knowledge of sucrose chemistry has picked up such momentum that even more growth is projected for the next decade. To further promote sucrochemical development, it has been proposed that the ISRF set up a center for the derivative samples that have been prepared (Jacques 1973). Much technical information on various sucrose compounds can be found in *Sucrose Chemicals* (Kollonitsch 1970).

A number of simple sucrose esters are easy to prepare and have already found their way into industrial use (Table 6.1). The ISRF still holds the patent right for the process to make sucrose esters. This process was originally designed to produce surfactants or edible detergents, but has since been found to produce a very effective bread conditioner. The use of this agent permits the enrichment of wheat flour with 20 to 30% protein components (such as fish-powder or soybean meal), and results in better volume and storage stability of the baked material. Products using sucrose esters have also been shown to enhance the growth rate of rats as a result of easier digestibility of the nutrients (Tsen *et al.* 1971).

Considerable potential for sucropolymers has been recognized in resins for surface coatings, textile treatment, and paper sizing; in plastics for molding and casting in adhesive agents; and in the manufacture of fiber and foamed elastomers. As yet, none of the experimentations has been developed to the production stage; the great technical difficulty in the preparation or production of sucropolymers for these specific purposes has already been discussed. In addition, the economic factors are unfavorable at this time.

TABLE 6.1

SUCROSE DERIVATIVES AND THEIR USES

Type of Derivative	Specific Compound	Commercial Uses
Acids	Arabonic acid	Chelating agent
	Gluconic acid	Cleaning agent
	Levulinic acid	Plasticizer
	Lactic acid	Plasticizer
	Oxalic acid	Textile chemical
Esters	Distearate	Emulsifier, bread conditioner
	Diisobutyrate	Plasticizer
	Hexaacetate	Viscosity modifier
	Hexalinoleate	Emulsifier, paint resin
	Monoacetate	Humectant
	Monomethacrylate	Resin monomer
	Monostearate	Surfactant, bread conditioner
	Octaacetate	Denaturant, plasticizer
	Octabenzoate	Plasticizer
	Polycarbonate	Plastic monomer
Ethers	Heptaallyl	Drying oil
	Octadecyl	Surfactant
	Octacyanoethyl	Dielectric, wood stabilizer
	Octohydroxypropyl	Monomer for cross linking polymer foam
	Tetracarboxyethyl	Chelating agent
Acetals	Cetyloxyethyl	Surfactant
Urethanes	N-Alkyl Sulfonyl	Surfactant
Xanthates	S-Alkyl Monoxanthates	Surfactant, chelating agent

In reviewing the possible routes for developing sucrochemicals, Hickson (1971) listed 16 market outlets large enough to permit the dispensing of sizable amounts of sucrochemicals; half of these outlets have been eliminated either because of their incompatibility with the economic background or because the feasibility of the use of sucrose was low, as viewed from the properties of the sucrose molecule. Thus, Hickson rated animal feeds, explosives, fuels, elastomers, lubricants, and soil conditioners as having low priority. His outlook for sucrose in the development of fibers, adhesives, paper and textile chemicals, plastics and plasticizers, pesticides, surface coatings, and surfactants was much brighter.

The versatile modifications of sucro-based chemicals are often compared with petroleum-based chemicals. The price-volume ratio of sucrose is 2 to 3 times as high as that of ethylene, benzene, and butadiene of petroleum origin (Anon. 1971B). The large difference in cost between the two prospective competitors is the primary economic consideration against investment in the chemical utilization of sucrose. Contrary to this pessimistic view, enthusiastic advocates of sucrochemicals seek comfort from the fact that most sucrochemicals possess certain unique properties, e.g., biodegradability of polymers; price competition with petrochemicals would be of little concern when the kinds or quality of products are different and not comparable. Furthermore, the cost advantage of petrochemicals is not unshakable, as the underground resources are rapidly diminishing with costs rising accordingly. The supply of sugar can be replenished annually, and there is elasticity in the adjustment of supply and demand.

BIBLIOGRAPHY

ANON. 1971A. Sugar Report. No. 226. USDA. Washington, D.C.
ANON. 1971B. Growth sags for top 50 volume chemicals. Chem. Eng. News *49*, 20, 15-17.
BARKER, S. A. *et al.* 1963. Sucrose derivatives. J. Chem. Soc. pp. 3403-3406.
HICKSON, J. L. 1971. Utilization of Sucrose by Chemists. *In* Sugar. Yudkins, J., Edelman, J., and Hough, L. (Editors). Butterworths, London.
HOUGH, L. 1971. The Chemical Reactivity of Sucrose. *In* Sugar. Yudkin, J., Edelman, J., and Hough, L. (Editors). Butterworths, London.
JACQUES, P. 1973. Sugar Based Anti-Foaming Agents II, ISRF Symposium, Expansion of Sugar Uses Through Research. Bethesda, Md.
KOLLONITSCH, V. 1970. Sucrose Chemicals. ISRF Inc., Bethesda, Md.
TSEN, C. C., HOOVER, W. J., and PHILLIPS, D. 1971. High Protein Breads Bakers Dig. *45*, *2*, 20.

CHAPTER 7

John D. Commerford │ Corn Sweetener Industry

There are 12 corn wet-milling companies in the United States. These companies operate or have under construction a total of 17 plants, 16 of which have facilities for producing corn sweeteners (Table 7.1). Total annual production of corn sweeteners exceeds 4.5 billion lb. Corn refiners' shipments of syrup and dextrose since 1950 are shown in Table 7.2.

TABLE 7.1

CORN SWEETENER MANUFACTURERS

Company and Plant Location	Products	
	Corn Syrup	Dextrose
American Maize-Products Company Hammond, Indiana	X	
Amstar Corporation Dimmitt, Texas	X	
Anheuser-Busch, Inc. Lafayette, Indiana	X	
Cargill, Incorporated Cedar Rapids, Iowa Dayton, Ohio	X	
Clinton Corn Processing Company Clinton, Iowa	X	X
Corn Sweeteners, Inc. Cedar Rapids, Iowa	X	
CPC International Inc. Argo, Illinois Pekin, Illinois North Kansas City, Missouri Corpus Christi, Texas	X	X
Grain Processing Corporation Muscatine, Iowa	X	
The Hubinger Company Keokuk, Iowa	X	
Penick & Ford, Limited Cedar Rapids, Iowa	X	
A. E. Staley Manufacturing Company Decatur, Illinois Morrisville, Pennsylvania	X	X

TABLE 7.2

SHIPMENTS OF SYRUP AND DEXTROSE, 1950-1971

Calendar Year	Corn Syrup[1]	Dextrose[1]
1950	1,484,094	726,862
1955	1,580,334	657,643
1960	1,968,259	807,469
1965	2,961,519	1,031,522
1970	3,328,070	1,204,510
1971	3,530,101	1,265,386

Source: *Sugar Reports*, USDA, ASCS 1972.

[1] In 1000 lb units.

THE CORN WET-MILLING PROCESS

Corn wet-milling can be divided into three steps: steeping, milling, and product recovery. A schematic diagram of this process is shown in Fig. 7.1.

Shelled corn, either from storage or as received, is cleaned and transferred to large stainless steel or wooden steep tanks. It is at this point that the process becomes corn wet-milling. Water containing a small amount of sulfur dioxide to control fermentation and facilitate softening is circulated in a countercurrent manner through a series of steep tanks. Each tank contains from 2,000 to 6,000 bu of corn. The steepwater is maintained at about 120°F and steeping is carried out for about 40 hr, during which the kernels take up water and swell to their maximum size. The steepwater removes soluble components from the corn and loosens the starch from the protein matrix.

The steepwater is then drained off and the softened kernels are conveyed to degerminating mills, which are designed to tear apart the kernels, freeing the germ containing corn oil, and loosening the hull. The germ fraction is then separated by passing the stream of broken kernels through centrifugal hydrocyclones.

The remaining material goes through a second grinding step to free the starch and gluten from the coarse particles. A system of screens, filters and washing devices is used to separate the hulls and fibers. Starch is separated from gluten by high-speed centrifuges. The lighter gluten fraction is taken off and the underflow is a dilute slurry of corn starch. Several additional steps involving recycling of certain streams, further washing, and dewatering complete the process.

The germ is cleaned, dried, expelled and extracted to obtain crude corn oil, which is refined to obtain a clear, edible oil suitable for salad dressing and frying. The spent germ is marketed as corn germ meal or is combined with other materials in corn gluten feed or meal.

The hulls and fibers are pressed and become components of corn gluten feed.

The protein-rich corn gluten is dried and sold as such, or as corn gluten meal; it may also be combined with other products to become corn gluten feed.

Even the steepwater is recovered. Light steepwater is pumped to multiple-effect evaporators and concentrated to about 50% solids. It

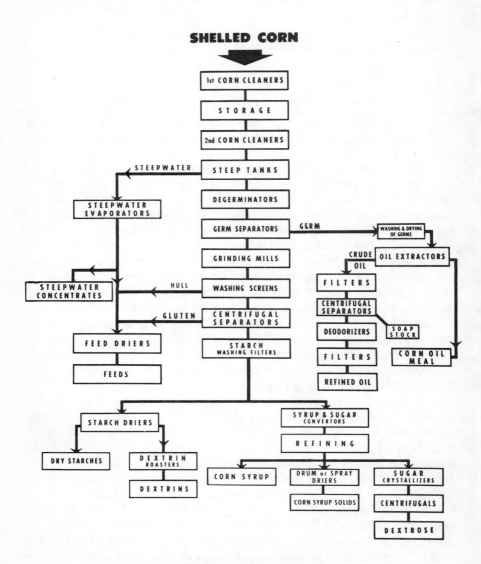

FIG. 7.1. THE CORN REFINING PROCESS

is used by the fermentation industry in the production of antibiotics, or is combined with other ingredients as an effective feed supplement, providing valuable nutrients.

CORN SWEETENER PRODUCTS

There are three main types of product: glucose syrup (corn syrup), maltodextrins, and dextrose. Glucose syrups are manufactured in liquid or dry form, and dextrose is manufactured as dextrose monohydrate or dextrose anhydrous.

The corn wet-milling industry has adopted and recommended for worldwide and domestic usage the following definitions for these products:

Glucose syrup is a purified concentrated solution of nutritive saccharides obtained from starch and having a D.E. of 20 or more. This definition applies to those nutritive sweeteners containing mono-, di-, and higher saccharides, regardless of the botanical source of the starch. In the U.S., since corn is the usual source, glucose syrup is usually called corn syrup. If most of the water is removed from glucose syrup (corn syrup), the product is called dried glucose syrup (dried corn syrup), or simply corn syrup solids.

Dried glucose syrups are made by spray-drying the concentrated solutions. One manufacturer makes a product by spray drying a highly converted starch hydrolyzate under conditions which form granules consisting of dextrose and corn syrup solids. This product has a D.E. of 95 or above.

Maltodextrins are purified concentrated solutions (or dry products made therefrom) of nutritive saccharides obtained from starch and having a D.E. of less than 20.

Dextrose monohydrate is purified and crystallized D-glucose containing one molecule of water of crystallization with each molecule of dextrose.

Dextrose anhydrous is purified and crystallized D-glucose without water of crystallization.

The corn sweetener industry has used the term dextrose equivalent (D.E.) for many years in describing syrups. The term dextrose equivalent is a measure of the reducing sugar content. The method commonly used for measuring D.E. in the industry is based on the (Lane and Eynon) volumetric alkaline copper technique (CRA. 1972).

D.E. is defined as total reducing sugars expressed as dextrose and calculated as a percentage of the total dry substance. In addition to the general definitions of glucose syrup and maltodextrins based on dextrose equivalent and source, there are several other classifications based on composition or solids content.

For statistical reporting purposes glucose syrups are divided into four types based on D.E.:

Type I	20 D.E. up to 38 D.E.
Type II	38 D.E. up to 58 D.E.
Type III	58 D.E. up to 73 D.E.
Type IV	73 D.E. and above

Glucose syrups are also characterized according to solids content. In practice this is done by measuring the Baumé using a calibrated hydrometer (145 modulus). Baumé is related to specific gravity by the following equation:

$$°Bé(60°F/60°F) = 145 - \frac{145}{\text{True Specific Gravity } (60°F/60°F)}$$

Since the saccharide composition of a syrup manufactured by a given process is substantially the same for a given D.E., the specific gravity can be used to determine dry substance by reference to a calibration curve or table.

Ordinarily, the measurement is made at 140°F and the observed is increased by 1 to provide an approximate value at 100°F. The calculation is as follows:

Commercial Baumé = (Observed Baumé at 140°F) +1

The Baumé of most commercial syrups is in the range of 43 to 45°, corresponding to dry substance values of 79 to 85% (CRA 1969).

A third means of characterizing glucose syrups is by saccharide composition. Depending on the process used (e.g., acid or enzyme) and the reaction conditions, a variety of products can be produced.

The traditional acid-converted syrups can be described by D.E. and the composition is essentially fixed. For example, the dextrose content of a 42 D.E. acid-converted corn syrup is about 18%, and for

TABLE 7.3

COMPOSITION OF REPRESENTATIVE CORN SYRUPS

Type of Conversion	Dextrose Equivalent	Saccharides, %							
		Mono-	Di-	Tri-	Tetra-	Penta-	Hexa-	Hepta-	Higher
Acid	30	10.4	9.3	8.6	8.2	7.2	6.0	5.2	45.1
Acid	42	18.5	13.9	11.6	9.9	8.4	6.6	5.7	25.4
Acid-enzyme	43	5.5	46.2	12.3	3.2	1.8	1.5	. . .	29.5[1]
Acid	54	29.7	17.8	13.2	9.6	7.3	5.3	4.3	12.8
Acid	60	36.2	19.5	13.2	8.7	6.3	4.4	3.2	8.5
Acid-enzyme	63	38.8	28.1	13.7	4.1	4.5	2.6	. . .	8.2[1]
Acid-enzyme	71	43.7	36.7	3.7	3.2	0.8	4.3	. . .	7.6[1]

[1] Includes heptasaccharides.

60 D.E. acid-converted corn syrup the dextrose content is about 36% (see Table 7.3). However, with the advent of acid-enzyme and dual-enzyme conversions, the exact composition of a syrup can no longer be predicted by D.E.

Instead the composition is usually stated in terms of D.E. and one or more characterizing features which indicate a specific product type or saccharide distribution. The composition of several of these corn syrups is also shown in Table 7.3.

One of the best examples of this is 43 D.E. high-maltose syrup made by an acid-enzyme process. The enzyme is selected to produce a high maltose level and minimal dextrose. Whereas regular 42 D.E. acid-converted syrups contained 18% dextrose and 14% maltose, high-maltose types may contain 5% dextrose or less, and 46% maltose.

The saccharide composition may be determined by quantitative paper chromatography or gas-liquid chromatography of the silyl derivatives.

Syrup Process

The starch slurry obtained by corn wet-milling contains about 4 lb of dry starch per gal. The starch is in the form of hydrated granules. In order to convert the starch from granule form to corn syrup, the granules must be gelatinized and the dispersed starch paste depolymerized. These steps are accomplished by subjecting the slurry to a combination of heat, shear, and hydrolytic actions under various conditions of time and temperature. In the acid-conversion process, the slurry is treated with a mineral acid to obtain a pH of about 2 and is pumped to a converter. The steam pressure is adjusted to achieve the desired temperature and the starch is simultaneously gelatinized and depolymerized. The process is terminated by neutralizing the converted syrup with a base such as soda ash. The resulting light liquor is clarified by filtration and/or centrifugation to remove suspended solid matter and residual fatty materials, and is concentrated by evaporation to about 60% solids.

The intermediate syrup is treated with powdered or granular carbon to further remove colored impurities. The syrup may also be ion-exchanged to remove soluble minerals and to lower the ash content. Final concentration is effected in large vacuum pans. The finished syrup is cooled and loaded into tank cars or trucks for shipment.

The acid-enzyme process is similar, except that the starch slurry is only partially converted to a given D.E. The light liquor is treated with an appropriate enzyme or combination of enzymes to complete the conversion. For example, in the production of the high-maltose syrup described previously, acid conversion is halted at a point where dextrose production is negligible. Then a maltose-producing enzyme such as the beta-amylase, obtained from malted barley, is added and the conversion continued under appropriate conditions to the desired level. The enzyme is then deactivated and the purification, clarification and concentration procedures continued as in acid-converted syrup production.

Dual-enzyme conversions are similar, except that after one treatment a second enzyme is added to complete the conversion.

In enzyme-enzyme processes, the starch granules are cooked and the preliminary starch splitting or depolymerization is brought about by an α-amylase rather than by means of acid. These processes have advantages in preparing specific syrups, since acid reversion products are not formed and higher molecular weight polysaccharides are the principal products. Of recent interest is the use of an iso-amylase to debranch the starch molecules at the 1,6-linkages. Japanese investigators have reported the use of this process as a starting point in the manufacture of maltose (Suzuki 1970).

Maltodextrin Process

Maltodextrins are produced by stopping the conversion process at an early stage, thus keeping the D.E. below 20. Both acid and enzyme processes can be used. After conversion to the desired level, the residual protein and fatty materials are removed by filtration, as in the refining of corn syrup. The light liquors are concentrated and are then clarified by carbon treatment prior to final concentration. Maltodextrins are spray-dried to give white, free-flowing powders, and are packed in multiwall bags.

Dextrose Process

The process for the manufacture of dextrose requires complete depolymerization of the starch substrate and recovery of the product by crystallization.

The starch slurry is gelatinized, as in the manufacture of corn syrup, and is partially converted by acid or α-amylase. Then a purified glucoamylase enzyme substantially free of transglucosylase activity is added to the hydrolyzate. When dextrose conversion is complete, the enzyme is deactivated, the dextrose liquor is filtered to remove residual suspended materials, and treated with granular or powdered carbon to remove trace impurities and color.

The liquor is concentrated to about 75% solids, cooled, and pumped into crystallizers. Crystallization is induced by seed crystals left in the crystallizer from the previous batch. The temperature is slowly lowered to about 25°C over a period of several days. Crystalline dextrose monohydrate precipitates from the mother liquor and is separated by centrifuges and washed while in the centrifuges with a spray of water. The wet crystals are then dried in warm air to about 8.5% moisture.

The mother liquor may be recycled or reconcentrated and a second crop of dextrose hydrate recovered.

Anhydrous dextrose is obtained by re-dissolving dextrose hydrate and refining the dextrose solution to obtain a highly purified clear filtrate. This solution is evaporated to a higher solids content, and anhydrous α-D-glucose is precipitated by crystallizing at an elevated temperature. The anhydrous crystals are separated by centrifugation, washed with a warm water spray, and dried.

Prior to the development of the enzyme technology dextrose was produced by an acid-conversion process.

TABLE 7.4

CORN SYRUP USES IN FOODS

Bakery products (bread, rolls, cakes, pies, cookies, icings)
Beverages (beer, ale, soft drinks)
Catsup
Canned fruits and vegetables
Confectionery
Desserts
Frozen foods
Ice cream and dairy products, including imitations
Infant foods
Jams, jellies and preserves
Prepared mixes
Syrups and toppings

TABLE 7.5

DEXTROSE USES IN FOODS

Bakery products (bread, rolls, pies, cookies, icings)
Beverages (beer, ale, soft drinks)
Canned fruits and vegetables
Catsup
Confectionery
Desserts
Frozen foods
Ice cream and dairy products
Infant foods
Intravenous feeding
Prepared mixes
Syrups

CORN SWEETENER USES

Corn sweeteners are used in practically all segments of the food industry. Some of the uses are shown in Tables 7.4 and 7.5. These uses are grouped into seven major classes. Table 7.6 shows the corn refiners' shipments of corn syrup and dextrose to those users for calendar year 1971.

TABLE 7.6

CORN REFINERS' SHIPMENTS
of
CORN SYRUP AND DEXTROSE
1971

| Type | Million Lb | |
of Buyer	Corn Syrup	Dextrose
Bakery, cereal and related products	587	464
Confectionery and related products	1,017	125
Ice-cream and dairy products	515	17
Beverages	252	32
Canners and packers, jams, jellies and preserves	407	77
Multiple and all other food uses	627	359
Nonfood uses	124	193
TOTALS	3,530	1,265

Source: *Sugar Reports*, USDA, ASCS 1972.

BIBLIOGRAPHY

CRA. 1972. Method E-26, Dextrose Equivalent, 4th Ed., Standard Analytical Methods of the Member Companies of the Corn Industries Research Foundation, Div. of Corn Refiners Assoc., Inc., Washington, D.C.

CRA. 1969. Critical Data Tables of the Corn Refiners Association, Inc., 1969 Edition, based on Fauser, E. E., Cleland, J. E., Evans, J. W., and Fetzer, W.,R., 1943, Baume—Dextrose equivalent—dry substance tables for corn syrup and corn sugar. Ind. Eng. Chem., Anal. Ed. *15*, 193.

SUZUKI, S., 1970, Novel industrial processes for enzymic conversion of starch. *In* Proceedings, Third International Congress, Food Science and Technology, G.,F. Stewart, and C. L. Willey (Editors). Inst. Food Technol., Chicago, Ill.

USDA. ASCS. 1972. Sugar Reports. No. 238, U.S. Dept. Agr., Washington, D.C.

CHAPTER 8

J. M. Newton
E. K. Wardrip

High-Fructose Corn Syrup

Since the first report of the production of sweet starch hydrolyzates by Kirchoff in 1812, the producers of starch hydrolyzates have had as a major objective the development of processes to increase the sweetness of starch syrups. Prior to 1938, the only progress made toward this objective was an increase in the degree of conversion as measured by dextrose equivalent (D.E.). All such commercial syrups were produced by acid hydrolysis, except for small quantities of malt-hydrolyzed starch syrups called "malt syrups." The acid-converted syrups were limited in their dextrose content and sweetness because conversions beyond 56 to 58 D.E. had distinct bitter flavors which could not be removed by the refining techniques available.

The first major improvement introduced by the corn wet-milling industry was the development of the acid-enzyme dual conversion process of Dale and Langlois (1940). In this process, the cornstarch is acid-converted to approximately 52 D.E., neutralized, and further converted with an amylolytic system producing maltose, primarily. These syrups have a D.E. of 61 to 65, are appreciably sweeter, do not have the conversion flavors, and do not crystallize as readily as a comparable D.E. acid-converted corn syrup.

The development of fermentation processes yielding gluco-amylases was commercialized in the early 1950's. These enzymes are utilized by the corn wet-milling industry to produce high-dextrose corn syrups with a D.E. range of 95 to 97, and containing up to 95% glucose (dextrose). Such syrups are appreciably sweeter than other corn syrups, but, because of their high dextrose content, they crystallize readily. However, by shipping and storing at elevated temperatures (130-140°F), the industry is able to deliver this sweeter, highly fermentable, nutritive corn syrup to food processors.

For many years the corn wet-milling industry has directed considerable research effort toward the development of corn syrups containing sufficient fructose to increase the sweetness of the syrup appreciably. Extensive investigations utilizing alkaline isomerization of dextrose-rich hydrolyzates have been undertaken. This process has not been successfully commercialized because of color, flavor, and composition problems.

The possibility of an enzyme process for isomerizing glucose to fructose has been recognized for some time. Marshall and Kooi (1957) reported the isolation of such an enzyme; a patent on this process was issued to Marshall (1960). In the same year, Tsumura and Sato (1960) reported the isolation of a specific soil organism that would produce an enzyme that converted glucose to fructose. Later, Tsumura *et al.* (1965) published on the production of an enzyme from *Streptomyces phaeochromogenes* that would isomerize glucose to fructose. The use of *Streptomyces* to produce glucose isomerase and the utilization of the enzyme was extensively discussed by Takasaki (1966).

Although numerous processes have been discussed in the literature, the basic process for producing high-fructose corn syrups in the U.S. is described in U.S. Patent 3,616,221 issued to Takasaki and Tanabe (1971). This patent is assigned to the Agency of Industrial Science and Technology, Tokyo, Japan. This agency has granted Standard Brands Incorporated an exclusive U.S. license, with the right to sub-license.

Several species of *Streptomyces* are identified by Takasaki and Tanabe (1971) as producing a glucose-isomerizing enzyme. The morphological characteristics of the preferred species are described, as well as culture media, fermentation conditions, and efficacy of the resulting enzyme in isomerizing dextrose to fructose. The isomerization of solutions of crystalline dextrose, hydrol, and high-D.E. corn syrups are also described.

An additional patent issued to Cotter *et al.* (1971) identifies the preferred organisms as *Streptomyces* sp. ATCC21175 and ATCC 21176. Taxonomical characteristics of *Streptomyces* sp. ATCC21175 are described in this patent. In addition to describing in detail the culturing of the organism and the application of the resulting isomerase to various starch hydrolyzates, this patent introduces the unique concept of carrying out the isomerization in the presence of various water-soluble sulfite salts. The water-soluble salts of sulfurous acid not only reduce the formation of color bodies, but also act as an enzyme stabilizer and as a buffer during the isomerization. This, of course, simplifies the refining of the isomerized syrups to yield water-white corn syrups containing approximately 42% fructose.

An excellent discussion of the research studies on glucose isomerase was presented by Takasaki *et al.* (1969). Using a *Streptomyces albus* culture isolated from soil, and cultured in a medium of 3% wheat bran, 2% corn steep liquor, and 0.024% $CoCl_2 \cdot 6H_2O$ at pH 7 and 30°C, they were able to produce sizable quantities of glucose isomerase. This procedure utilizes the earlier findings of Takasaki (1966) in which xylan-containing materials can be substituted for the more expensive xylose. These authors reported extensively on the physical and chemical properties of purified and crystallized glucose isomerase from *Streptomyces albus.*

A significant part of their report (Takasaki *et al.* 1969) is a description of a commercial process for isomerizing glucose. In their process the cells from 5 cubic meters of culture fluid were added to 30 tons of a solution containing 50% glucose, $0.005M$ $MgSO_4$, and $0.001M$ $CoCl_2$. The pH was maintained at approximately 7 with NaOH and the temperature at 65 to 70°C. After about 3 days the reaction mixture was filtered, purified with active carbon and ion-exchange resins, and concentrated to 75% solids. The finished high-fructose corn syrups contained 45 to 50% fructose and 50 to 55% glucose.

A most interesting innovation suggested by Takasaki *et al.* (1969) allows the enzyme to be used repeatedly. The glucose isomerase produced by *Streptomyces albus* was found to be predominantly intracellular. Heating the cells to over 60°C for about 10 min, fixed the enzyme within the cells. When these heat treated cells were used to isomerize a dextrose solution, the cells containing the enzyme could be recovered by centrifugation after the reaction was completed. The cells can be reused to isomerize another batch of dextrose solution; seven reuses were reported as being feasible.

One of the major problems in utilizing a bacterial fermentation for the production of enzymes is to obtain a sufficient quantity of enzyme in the fermentation vats to make commercial operations economically feasible. Since glucose isomerase is primarily produced intracellularly by microorganisms, and because whole cells are commonly used for the isomerization reaction, the highest possible enzyme-cellular material ratios are essential. Bengston and Lamm (1972) describe a process for obtaining increased yields of glucose isomerase from the *Streptomyces* genus. In this process, toxic agents were added to the viable microorganisms so that 90 to 95% of the microorganisms were inactivated. The term "microorganism" in this patent included both spores and vegetative cells. Actively growing *Streptomyces* cultures were treated with such chemicals as ethylenimine, hydrogen peroxide, 8-ethoxycaffeine, as well as by radiation from radioactive isotopes and ultraviolet light. The surviving cells were isolated, cultured, and checked for their ability to produce isomerase. These investigators claim that an increase of 30 to 50% in the yield of glucose isomerase can be obtained from colonies isolated from treated cultures. They also report that the cellular material produced in commercial fermentations using these isolates filters appreciably faster than that obtained from the parent culture. This is a definite commercial advantage.

A process for increased yields of glucose isomerase from the *Streptomyces* genus was described by Dworschack and Lamm (1972). In their process, a microorganism of the genus *Streptomyces*, which will assimilate xylan to form a glucose isomerase, was used as the microorganism. Their process disclosed that the use of aqueous dispersible material in the initial growth stages induced the microorganism to grow in a filamentous form. Various water-dispersible materials such as agar, carboxymethyl cellulose, and diatomaceous earth were used. Three- to fourfold increases in glucose isomerase yields were obtained by this process. This markedly increased the quantity of glucose isomerase that could be isolated from a fermenter and had a distinct effect on the economics of the commercial process for making high-fructose corn syrup.

Lloyd *et al.* (1972) described an advanced process for enzymatically converting a portion of the glucose in a solution to fructose. This patent stressed the ability to produce a syrup with minimal quantities of by-products that contribute to flavor, color, and odor. The method "provides a continuous, commercial process for enzymatically isomerizing glucose to fructose", as well as providing maximum "utilization of the glucose isomerase" and a final "glucose fructose syrup containing minimal color, ash and psicose."

One specific example described in the patent utilized *Streptomyces* ATCC21175 grown under submerged aerobic conditions. The finished fermenter broth was adjusted to 7.5 pH; the broth was heated in 30 min to 75°C and maintained at that temperature for 5 min. Filter aid (3%) was added to the broth, agitated, the broth filtered on a rotary drum filter precoated with diatomaceous earth, and the cellular material washed with water. The resulting filter cake was dried in a forced-air dryer (air temperature 49°C) for about 3.5 hr to yield a fixed isomerase preparation.

The fixed isomerase cake (256 lb) was slurried with a 65°C glucose-containing refined cornstarch hydrolyzate (60 gm glucose per 100 ml) at a pH of 7.0, 0.001 mole $CoCl_2$ per l, 0.1 mole $MgSO_4$ per l, and 0.006 mole $NaHSO_3$ per l. This slurry was pumped through a pressure leaf filter (6 leaves) until each leaf was coated with a 1.0 to 1.5 in. layer of the fixed isomerase preparation. The filter leaves had a total surface of 74 ft. Glucose solution was then pumped continuously through the filter at a rate such that 45% of the glucose was converted to fructose. The flow rate was gradually reduced as the activity of the fixed isomerase on the leaves diminished, so that the conversion of glucose to fructose remained at 45%. After 40 hr the flow rate through the leaves was 1.0 gal per min, and after 200 hr it was 0.5 gal per min. The glucose-containing solution had a color of 0.01 color unit as it entered the leaves and 0.02 color unit at the exit.

This patent reported isomerization of refined glucose-containing corn syrup hydrolyzate solutions to yield products containing from 30 to 60% glucose, 10 to 54% fructose, 0 to 30% other saccharides, and 0 to 1.0% psicose. This allowed for a series of products with varying nutritive saccharide composition.

Kooi and Smith (1972) described a process which utilizes a strain of *Streptomyces olivochromogenes* to produce glucose isomerase. The whole cells were collected on a food-grade inert carrier, washed, and dried. Application of this enzyme cake to a dextrose solution under controlled conditions of pH, time, and temperature isomerized about 45% of the glucose to fructose. The resulting isomerized syrup was then refined by typical processes applied to corn syrup and concentrated to 71% solids. The composition of the finished product was comparable to the commercially available products previously described, except that there was a small increase in the fructose and glucose content, with a concomitant reduction in other saccharides. The authors state that "beverage application studies made in the course of product development work indicate that this new syrup made from dextrose has functional utility in soft drinks."

Lee *et al.* (1972) described a process for producing glucose isomerase utilizing microorganisms from the genus *Arthrobacter*. Three of the strains evaluated were unique in that they produced glucose isomerase in the complete absence of xylose and xylan. By applying either the whole cells or cell-free extracts to starch hydrolyzates with a 95 D.E., syrups were obtained containing from 38 to 45% fructose.

A detailed description of the preparation of a crystalline glucose isomerase was presented by Yamanaka (1968). D-Xylose, D-glucose, and D-ribose were isomerized by this crystalline enzyme. The enzyme was produced by cultures of *Lactobacillus brevis*. This article is of particular interest because it refers to 32 publications on the production of glucose isomerase by a wide range of organisms, such as *Pseudomonas, Aerobacter, Escherichia, Bacillus, Brevibacterium, Paralactobacterium, Leuconostoc*, and *Streptomyces* species.

COMMERCIAL UTILIZATION OF IMMOBILIZED ENZYMES

The American Chemical Society in "Chemicals in 1972" (Anon. 1973) emphasizes the industrial importance of enzymes bound to insoluble carriers. They refer to the use by the corn wet-milling industry of "bound enzymes" to make fructose-containing nutritive saccharides from cornstarch.

Two commercial plants to manufacture high-fructose corn syrups have been built and are in commercial operation, as follows: Clinton Corn Processing Co. (A Division of Standard Brands Incorporated; Clinton, Ia.) and A. E. Staley Manufacturing Co. (Decatur, Ill.; Plant located at Morrisville, Pa.). Plans to build one additional plant to manufacture a similar fructose-rich syrup from cornstarch have recently been announced by CPC International, Englewood Cliffs, N.J.

The plant at Clinton, Ia. is the world's largest plant for producing high-fructose corn syrup. The principal product is a water-white solution of nutritive saccharides containing 50% dextrose and 42% fructose. On a solids basis it is equivalent to sucrose in sweetening value. The process utilizes an immobilized glucose isomerase produced by *Streptomyces* species. The immobilized isomerase is placed in reactors and a dextrose-containing solution is continuously pumped through the reactors. The massive concentration of glucose isomerase rapidly isomerizes glucose to fructose. After isomerization, the solution of nutritive saccharides is refined by conventional processes, utilizing filtration, carbon decolorization, ion-exchange deionization, and concentration to the desired solids content.

Because the enzyme is immobilized, it remains in a fixed position and isomerizes the soluble glucose as the solutions filter through the reaction beds. The immobilized enzyme has an extremely long life (hundreds of hours), and can be removed from the system by simply isolating a single reactor. The reactor can then be recharged and returned to the reactor system. Careful control of such variables as pH, temperature, rate of feed, etc. yields a process conducive to automatic control.

The overall process which involves continuous, semicontinuous, and batch operations, has been tied together by an on-line process control system reported by Harden (1972, 1973). The system requires both analog and digital, pneumatic and electronic equipment and techniques. The process involves liquefying raw cornstarch, saccharifying to glucose, refining, isomerizing the glucose to fructose, and again refining and concentrating the refined corn syrup. The ability to control a multienzyme, multiunit process manufacturing operation by automatic digital sequencing, combined with analog design for modulating controls, is a modern engineering success story. This is best demonstrated by Harden when he asks, "Does it work?" His answer is a simple, "Yes."

PRODUCTION AND USE OF HIGH FRUCTOSE CORN SYRUP

High-fructose corn syrups were first produced in limited commercial quantities in Japan in the mid-1960's. The first rail tank car commercial shipment in the United States was made in 1967. The original commercial products contained 14 to 16% fructose. Technological developments of the past 5.5 yr now make it possible to produce commercially products containing 47% fructose. Extension to higher fructose levels has been demonstrated and can be reduced to commercial practice as soon as the markets justify.

The principal high-fructose corn syrup (HFCS #1) being sold commercially in the United States today has the following typical analysis: moisture, 29%; dry substance, 71%; dextrose, 50% D.S.; fructose, 42% D.S.; other saccharides, 8% D.S.; ash, 0.03% D.S.; and nitrogen, 0.002% D.S.

Two additional high-fructose corn syrups have been offered. The composition of these are given in Table 8.1.

TABLE 8.1

COMPOSITION OF HFCS # 2 AND #3

	HFCS #2	HFCS #3
Moisture, %	29	29
Dry substance, %	71	71
Dextrose, % D.S.	51	54
Fructose, % D.S.	47	45
Other saccharides, % D.S.	2	1

Because HFCS #1 containing 50% glucose and 42% fructose has been sold commercially for some time, while HFCS # 2 and HFCS #3 were not offered until late 1972, the following comments on uses pertain to experience obtained from the sale and use of the first product (specifically, ISOMEROSE® 100 Brand High Fructose Corn Syrup).

High-fructose corn syrup has high sweetness, high fermentability, high humectancy, low viscosity, is water-white in color, and has a clean, nonmasking taste. With its low viscosity and reduced tendency to crystallize, it is easier to ship, store, and blend than other corn syrups. When maintained at slightly elevated temperatures (80 to 90°F) it can be pumped easily and stored for extended periods without discoloration, crystallization, or fermentation.

High-fructose corn syrup, on a dry substance basis, has the same sweetness as an equivalent weight of sucrose. Therefore, it can be used in food products on an equivalent sweetness basis, except in those products where a dry sweetener is required. High-fructose corn syrup is available only in liquid form.

Carbonated beverages in all flavors are being produced commercially using high-fructose corn syrup as part or all of the sweetener. They are equivalent in all respects to beverages produced from sucrose or invert syrup.

Fountain syrups and flavor concentrates are being manufactured using high-fructose corn syrup as the total sweetener. Since these products, when diluted with carbonated water, are comparable to the carbonated beverages previously described, the same satisfactory results are obtained. The syrups and flavor concentrates are also used to produce still beverages. Here the high-fructose corn syrup may be used in conjunction with dextrose and high-dextrose equivalent corn syrups. Such factors as sweetness, flavor, storage stability, pH, color, and microbiological profile are equivalent to or better than previous commercial formulas.

Pickle products, including sweet and candied sweet pickles, sweet pickle relish, and related pickle products are produced with high-fructose corn syrup. Sweet pickles require a sweetener that easily penetrates the cell membranes of the pickle. The 92% monosaccharide content of high-fructose corn syrup satisfies both the sweetener and penetration requirements. The color, crispness, flavor, and overall appearance of the pickles are excellent.

Catsup utilizing a mixture of high-D.E. corn syrup and high-fructose corn syrup is produced in numerous commercial operations. The bulk handling of all liquid sweeteners has resulted in major production economies. The product exceeds all quality standards and processing problems have been minimized.

High-fructose corn syrups alone or in conjunction with other corn syrups are used commercially as the effective sweetener for pourable and spoonable salad dressings. Sweetness, flavor, viscosity, emulsion stability, body, and texture are maintained at the desired standard.

Because yeast-raised baked goods require high fermentability and a residual sweetness, high-fructose corn syrup alone or in conjunction with other corn syrups becomes an ideal sweetener for this bakery application. The fructose content also contributes to the desirable brown color of all types of baked goods. Additional baked goods utilizing high-fructose corn syrup include sweet goods, cakes, fig bars, cookies, and other miscellaneous products. Other commercial applications include table syrups, apple sauce, confections, toppings, maraschino cherries, pie fillings, ice cream, and frozen specialties.

RESEARCH AND THE FUTURE

The National Science Foundation (NSF), starting in 1972, established a broad research program to exploit enzyme catalysts for wide usage in industrial processes. Historically, even though enzymes have been shown to be specific and efficient catalysts for chemical reactions, their industrial usage has been limited because of the lengthy and costly isolation procedures, the relative instability of the active enzyme, and the single use of the enzyme in aqueous systems.

Immobilization of enzymes on insoluble carriers has been demonstrated by many investigators. Generally, such immobilized enzyme preparations have greater stability which permits their use in continuous processes. To expedite the collection of information to assist in this development, NSF funded research projects totaling $1,950,000 in 1972, and proposed a funding of $2,400,000 for 1973. Of the 21 projects funded in 1972, 19 were specifically directed toward some aspect of the immobilization of various enzymes on many different carriers. At least four of the projects have as their objective the development of processes similar to or competitive with the commercial process now being utilized to manufacture high-fructose corn syrup.

The NSF program includes funding for "An Information Exchange in Enzyme Technology". The first two issues of the resulting publication, present an extensive summary of available literature references on enzyme immobilized on or in water-insoluble carriers (Hultin *et al.* 1972A, 1972B).

Additional issues of Enzyme Technology Digest (Anon. 1972) propose to publish up-to-date information on this rapidly evolving field of research. There, of course, continues to be a high degree of interest in immobilized enzymes in various industrial research laboratories. The developments generated from those studies will be reported primarily in the patent literature.

BIBLIOGRAPHY

ANON. 1973. Chemicals in 1972. Am. Chem. Soc., Washington, D.C.
ANON. 1972. Enzyme Technol. Dig., NEUS Inc., P.O. Box 1365, Santa Monica, Cal.
BENGSTON, D., and LAMM, W. R. 1972. Process for isomerizing glucose to fructose. U.S. Patent 3,654,800. Apr. 4.
COTTER, W. P., LLOYD, N. E., and HINMAN, C. W. 1971. Method for isomerizing glucose syrups. U.S. Patent 3,623,953. Nov. 30.
DALE, J. K., and LANGLOIS, D. P. 1940. Syrup and method of making the same. U.S. Patent 2,201,609. May 21.
DWORSCHACK, R. G., and LAMM, W. R. 1972. Process for growing microorganisms. U.S. Patent 3,666,628. May 30.

HARDEN, J. D. 1972. On-Line control optimizes processing, Part I. Food Eng., Dec., 59-62.

HARDEN, J. D. 1973. On-Line control optimizes processing, Part II. Food Eng., Jan., 65-67.

HULTIN, H. O., KITTRELL, J. R., and LAURENCE, R. L. 1972. Enzymes and reactor engineering. Enzyme Technol. Dig. 1, 1, 15-33.

HULTIN, H. O., KITTRELL, J. R., and LAURENCE, R. L. 1972. Enzymes and reactor engineering. Enzyme Technol. Dig. 1, 2, 46-71.

KOOI, E. R., and SMITH, R. J. 1972. Newest natural sweetener: dextrose-levulose syrup from dextrose. Food Technol. 26, 9, 57-59.

LEE, C. K., HAYES, L. E., and LONG, M. E. 1972. Process of preparing glucose isomerase. U.S. Patent 3,645,848. Feb. 29.

LLOYD, N. E., LEWIS, L. T., LOGAN, R. M., and PATEL, D. N. 1972. Process for isomerizing glucose to fructose. U.S. Patent 3,694,134. Sept. 26.

MARSHALL, R. O. 1960. Enzymatic process. U.S. Patent 2,950,228. Aug. 23.

MARSHALL, R. O., and KOOI, E. R. 1957. Enzymic inversion of D-glucose to D-fructose. Science 125, 648-649.

TAKASAKI, Y. 1966. Studies on sugar-isomerizing enzyme. Production and utilization of glucose isomerase from Streptomyces sp. Agr. Biol. Chem. 30, 1247-1253.

TAKASAKI, Y., KOSUGI, Y., and KANBAYASHI, A. 1969. Streptomyces Glucose Isomerase. Fermentation Advances, Academic Press, 561-589.

TAKASAKI, Y., and TANABE, O. 1971. Enzyma method for converting glucose in glucose syrup to fructose. U.S. Patent 3,616,221. Oct. 26.

TSUMURA, N., SAKAKURA, E., ISHIKAWA, M., and SATO, T., 1965. Enzymatic conversion of D-glucose to D-fructose. Hakko Kyokashi 23, 516-520.

TSUMURA, N., and SATO, T. 1960. Conversion of D-glucose to D-fructose by a strain of soil bacteria. Bull. Agric. Chem. Soc. Japan 24, 326-327.

YAMANAKA, K. 1968. Purification, crystallization and properties of the D-xylose isomerase from Lactobacillus brevis. Yamanaka, K. Biochim. Biophys. Acta 151, 670-680.

CHAPTER 9

G. A. Brooks
M. O. Warnecke
J. E. Long

Sweetness and Sensory Properties of Dextrose-Levulose Syrup

It is now commercially feasible to produce a Dextrose-Levulose Syrup (DLS) by the enzymatic isomerization of dextrose (Kooi and Smith 1972). This new sweetener is a purified aqueous solution containing 45% levulose. A comparison of the analysis of DLS with medium and total invert from sucrose is illustrated in Table 9.1. As the table shows, the DLS composition is similar to that of total invert.

TABLE 9.1

TYPICAL ANALYSES OF SYRUPS

Analysis	Dextrose-Levulose	Sucrose Invert	
		Medium	Total
Total solids, %	71	76	72
Levulose, %[1]	45	28	44
Dextrose, %[1]	54	29	47
Sucrose, %[1]	0	40	6
Polysaccharides, %[1]	1	3	3

[1] Dry basis.

The objective of the present work was to determine the suitability of DLS in soft drinks. Initial studies were made in two aqueous model systems. One model involved only distilled water; the other was acidified and carbonated. In the third phase, DLS was tested in commercially formulated soft drinks. Each system was evaluated by taste panels to determine comparative sweetness.

The literature contains much information on the relative sweetness of levulose, dextrose and sucrose as individual sweeteners and in combinations. Assigning sucrose a sweetness level of 100, dextrose solutions up to 15% concentration have been reported to have a relative sweetness of 53 to 87 (Renner 1939; Dahlberg and Penczek 1941). At concentrations of 40 to 50%, the relative sweetness of dextrose has been reported as high as 100 (Dahlberg and Penczek 1941; Eickelberg 1940). Values in the range of 79 to 180 have been given as the relative sweetness of levulose in comparison to sucrose (Tsuzuki and Yamazaki 1953; Redfern and Hickenbottom 1972).

Several values for the relative sweetness of dextrose and levulose between these extremes have been reported (Biester *et al.* 1925; Cameron 1943, 1944, 1945; Deer 1922; Eickelberg 1940; Lawrence and Ferguson 1959; Lichtenstein 1948; Nieman 1960; Pangborn 1963; Whymper 1955; Yamaguchi 1970A, 1970B). A high-fructose corn syrup at 11 to 17% solids was reported about as sweet as sucrose (Wardrip 1971).

Much of the variation in the reported values results from the test methods used, concentrations of the sweetener, stereochemical configuration, and the organoleptic responses of the subjects (Cameron 1947; Moskowitz 1970; Pangborn and Gee 1961; Schultz and Pilgrim 1957). Interactions with other chemical components such as salt, acids, and flavorings also affect perceived sweetness (Amerine *et al.* 1965; Beebe-Center *et al.* 1959; Berg *et al.* 1955; Fabian and Blum 1943; Kamen *et al.* 1960; Pangborn 1960, 1961, 1962).

Relative sweetness values, in general, have been determined in model solutions rather than in formulated food systems. Problems may be encountered when transposing sweetness results from aqueous systems to food products because the interactions of flavor components and texture in the food product may influence sweetness perception (Mackey 1958; Mackey and Jones 1954; Mackey and Valassi 1956; Stone and Oliver 1966; Vaisey *et al.* 1969). Nevertheless, evaluation in aqueous solutions is a valuable screening process to give direction as to levels in a final product.

EVALUATION

Procedure

A panel was selected by testing 96 persons from the research laboratories for their ability to distinguish 1% differences in sucrose solutions over a range of 8% to 12% solids, the normal concentration for soft drinks. From these candidates the best panel, consisting of 24 men and 2 women ranging in age from 25 to 50 years, was selected. Twenty members from this group were used for each evaluation.

The paired comparison test was chosen as being the most discriminating in meeting the objectives of the study. This was in agreement with results reported by Byer and Abrams (1955). The only question asked of the panelists was "Which sample is sweeter?". However, comments were often volunteered by the panelists regarding flavor characteristics other than sweetness. No more than two pairs were served at each sitting. Forty-ml samples were served at 4 to 10°C.

In these experiments, the DLS described by Kooi and Smith (1972) was used. This syrup in separate experiments was shown to be fully equivalent in flavor and sweetness to a corresponding blend made from crystalline dextrose and crystalline levulose. The commercial medium inverts used in this study contained 45 to 50% sucrose. In commerce the sucrose level in medium invert may range from 30 to 60%. All blends of DLS and sucrose were made on a dry solids basis.

Samples were considered significantly different when a confidence level of 95% or higher was obtained. With a 20 member panel, using the paired comparison test, 15 agreeing judgments constitutent a 95% confidence level. In the tables, when a comparison is shown to be significantly different, the confidence levels include 95% and higher.

Evaluation in Distilled Water

Sugar solutions were prepared on a w/w basis and held overnight at 3°C before testing. The solids level of the solutions were determined using an Abbe precision refractometer and the appropriate refractive index tables.

Comparison of Sweetness of DLS with Sucrose and Medium Invert.—DLS was compared to sucrose for sweetness at concentrations of 10 to 25% solids. The results are shown in Table 9.2. At

TABLE 9.2

COMPARATIVE SWEETNESS OF DEXTROSE-LEVULOSE SYRUP (DLS) AND SUCROSE (S) IN DISTILLED WATER

Sweetener	% Solids	Sweetness Response[1]
S	10	*Sweeter
DLS	10	—
S	11.5	N.S.
DLS	11.5	
S	14	N.S.
DLS	14	
S	16	N.S.
DLS	16	
S	19	N.S.
DLS	19	
S	25	N.S.
DLS	25	

[1] 20 panelists
N.S. Not significant
* p = .05

TABLE 9.3

COMPARATIVE SWEETNESS OF DEXTROSE-LEVULOSE SYRUP (DLS) AT VARYING
SOLIDS AND SUCROSE (S) IN DISTILLED WATER

Sweetener	% Solids	Sweetness Response[1]
S	10	*Sweeter
DLS	10	—
S	10	
DLS	10.5	N.S.
S	10	
DLS	11	N.S.

[1] 20 panelists
N.S. Not significant
* p = .05

TABLE 9.4

COMPARATIVE SWEETNESS OF DEXTROSE-LEVULOSE SYRUP (DLS) AND
MEDIUM INVERT (MI) IN DISTILLED WATER

Sweetener	% Solids	Sweetness Response[1]
MI	10.25	*Sweeter
DLS	10.5	—
MI	11.8	*Sweeter
DLS	12.0	—
MI	16	*Sweeter
DLS	16	—
MI	19	*Sweeter
DLS	19	—
MI	25	
DLS	25	N.S.

[1] 20 panelists
N.S. Not significant
* p = .05

10% solids (concentration), the sucrose was significantly sweeter.
However, at 11.5% and higher, there was no statistically significant
difference in discernible sweetness. To determine the overage of DLS
needed to match the sweetness of sucrose the comparisons in Table
9.3 were made. At a 10% sucrose solids, a 5% overage of DLS was
needed to match sucrose. Table 9.4 presents the comparison of DLS
to medium invert over a total solids range of 10 to 25%. Only at the
25% solids level was there no significance sweetness difference
between the two.

Higher overages of DLS were needed to match the sweetness of medium invert than were needed to match sucrose. Table 9.5 shows, on the average, that 10 to 12% higher concentrations of DLS were needed to match the sweetness of medium invert when the latter was present at 10% solids. Medium invert at 11.8% required a 6% overage of the DLS for equivalent sweetness. This is an initial difference and it decreases in acidic systems as the sucrose inverts. This will be discussed in Fig. 9.1 and Table 9.9

TABLE 9.5

COMPARATIVE SWEETNESS OF DEXTROSE-LEVULOSE SYRUP (DLS) AT VARYING SOLIDS AND MEDIUM INVERT (MI) IN DISTILLED WATER

Sweetener	% Solids	Sweetness Response[1]
MI	10.25	*Sweeter
DLS	11.0	—
MI	10.25	N.S.
DLS	11.5	
MI	11.8	*Sweeter
DLS	12.0	—
MI	11.8	N.S.
DLS	12.5	

[1] 20 panelists
N.S. Not significant
* p = .05

Comparison of the data of DLS vs sucrose and medium invert leads to the conclusion that, at the same solids, medium invert is sweeter than sucrose.

Synergism of Sucrose-Dextrose-Levulose.—In an acidic environment sucrose inverts into a blend of sucrose, dextrose and levulose. This carbohydrate system will have a different sweetness impression from that of sucrose alone. Cameron (1947) and Stone and Oliver (1969), among others, have described a synergistic sweetness resulting from combinations of sucrose, dextrose and levulose relative to sucrose alone.

DLS exists as monosaccharides and, therefore, remains unchanged in acidic systems.

Experiments were performed to determine the level of sucrose needed in DLS to give a blend having sweetness equal to sucrose and to medium invert at equal concentrations of total sweetener (Tables 9.6 and 9.7). A 5% addition of sucrose to DLS increased the sweetness of the blend to equal that of sucrose. All blends with 15%

or more sucrose were significantly sweeter than sucrose alone. A 25% or higher concentration of sucrose in DLS matched the sweetness of medium invert.

TABLE 9.6

COMPARATIVE SWEETNESS OF DEXTROSE-LEVULOSE SYRUP (DLS) AND SUCROSE BLENDS AND SUCROSE (S) IN DISTILLED WATER

Comparison	% Solids	Sweetness Response[1]
S	10	*Sweeter
DLS	10	—
S	10	
5% S/95% DLS	10	N.S.
S	10	—
15% S/85% DLS	10	*Sweeter
S	10	—
25% S/75% DLS	10	*Sweeter
S	10	—
50% S/50% DLS	10	*Sweeter
S	10	—
75% S/25% DLS	10	*Sweeter

[1] 20 panelists
N.S. Not significant
* p = .05

TABLE 9.7

COMPARATIVE SWEETNESS OF DEXTROSE-LEVULOSE SYRUPS (DLS)/SUCROSE (S) AND MEDIUM INVERT (MI) IN DISTILLED WATER

Sweetener	% Solids	Sweetness Response[1]
MI	10	*Sweeter
S	10	—
MI	10	*Sweeter
5% S/95% DLS	10	—
MI	10	*Sweeter
15% S/85% DLS	10	—
MI	10	
25% S/75% DLS	10	N.S.
MI	10	
50% S/50% DLS	10	N.S.
MI	10	
75% S/25% DLS	10	N.S.

[1] 20 panelists
N.S. Not significant
* p = .05

This further supports the previous conclusions that medium invert is sweeter than sucrose at the solids level found in soft drinks. From a practical standpoint the data show that a blend of 50% DLS and 50% sucrose has the same sweetness properties as commercial medium invert. In fact, the blend may need as little as 25% sucrose. The composition of the blend and medium invert would be the same when both contain the same amount of sucrose.

Evaluation in Acidic Carbonated Water Solutions

The next step was to evaluate the sweeteners in acidified, carbonated water. This demonstrates the effect of carbonation and acid on sweetness and it also resembles a soft drink, minus the flavor.

Studies of the effect of carbonation on sweetness were not found in the literature, but several reports have shown that acid decreases relative sweetness in aqueous solutions (Pangborn 1961; Spencer 1971; Stone *et al.* 1968). In addition, sucrose has been found to depress the sourness of acids (Pangborn 1960, 1961; Kamen *et al.* 1960).

Acidified sugar solutions (w/v basis) were added to bottles of carbonated water at 3°C after the appropriate amount of carbonated water had been decanted. The bottles were recapped and shaken to mix the components. The carbonated water was obtained from a local soft-drink bottler. Unless otherwise stated, phosphoric acid was used. All solutions were tasted within two days.

Sweetness of Sucrose During Inversion.—Sucrose will invert in an acid-carbonated beverage. Therefore, a carbonated beverage will contain varying amounts of sucrose, dextrose and levulose depending upon the degree of inversion. The first phase of this study indicated that this beverage could vary in sweetness. The relative sweetness of sucrose and totally inverted sucrose have been reported by Sale and Skinner (1922) and Willaman *et al.* (1925).

To determine the variation in sweetness during inversion, crystalline dextrose and crystalline levulose and sucrose were blended to simulate various stages of inversions with the weight gain added. These were compared to 10% sucrose. The results are shown in Fig. 9.1. As the sucrose "inverted", sweetness increased until about 50 to 85% inversion was reached. At about 25% remaining sucrose, the sweetness began to decrease and the final total invert, with 5% more solids, was equal in sweetness to sucrose. This again verified the synergism observed in a sucrose-dextrose-levulose mixture.

This same experiment can be used to illustrate what happens in an acidic aqueous solution containing medium invert. Initially it would have maximum sweetness, but this would decrease after about 75%

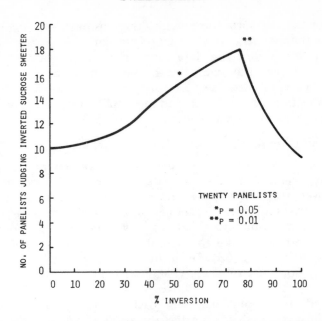

FIG. 9.1 SWEETNESS PATTERN OF SUCROSE DURING INVERSION

inversion. The sweetness would then approach that of sucrose or total invert.

DLS with Citric and Phosphoric Acid.—Several types of acids are used in carbonated beverages, citric acid and phosphoric acid being the most common. Citric acid is normally used in fruit-flavored drinks and phosphoric acid in colas. The effect of these acids on the sweetness and flavor of DLS was checked on a 75% DLS/25% sucrose blend. The control was medium invert. The data shown in Table 9.8 indicate that the sweetness of the blend was not affected by the acidulant. No flavor differences were noted by the panelists.

Storage in Acidic Carbonated Solutions.—Selected sweetener blends were prepared on the Q-Plus Bottling and Canning unit (Bohm and Hoynak 1967). In this pilot plant scale bottling machine, the sweetener solutions were mixed with phosphoric acid to a pH of 2.8 and carbonated to a level of 3.5 volumes. These bottled solutions resembled unflavored soft drinks. The solutions were evaluated for sweetness after 1 day, 1 month, and 3 months. Comments were requested on flavor, since this was the first long-term storage test.

The overall results (Table 9.9) were similar to those obtained in distilled water. However, it was a little more difficult for the panel to find differences in sweetness. The DLS at 5% higher concentration

TABLE 9.8

EFFECT OF PHOSPHORIC OR CITRIC ACID ON SWEETNESS IN CARBONATED WATER

Comparison	% Solids	pH	Sweetness Response[1]
MI	10	2.8	N.S.
25% S/75% DLS (0.1% Citric acid)	10	2.8	
MI	13	2.8	N.S.
25% S/75% DLS (0.1% Citric acid)	13	2.8	
MI	10	2.7	N.S.
25% S/75% DLS (0.02% Phosphoric acid)	10	2.7	
MI	13	2.7	N.S.
25% S/75% DLS (0.02% Phosphoric acid)	13	2.7	

[1] 20 panelists
N.S. Not significant

was equal to sucrose initially. After 1 and 3 months storage at room temperature, the sucrose solution was sweeter because of inversion. The DLS at 5% more solids was less sweet than medium invert initially, but at 1 and 3 months there was no significant difference. This was the result of decreased sucrose levels due to inversion. A 50/50 blend of DLS and sucrose was equivalent to medium invert at all test periods. All solutions prepared with DLS exhibited good flavor properties throughout the evaluation period.

Evaluation of Dextrose-Levulose Syrup in Commercial Soft Drinks

DLS is under extensive evaluation in commercial soft drinks. This includes ginger ale at approximately 9% sweetener solids, colas at about 11% solids and fruit flavors at about 14% solids. The results to date have been similar to those obtained in the model systems.

Two approaches to the use of DLS have been taken. The first is to replace the existing sweetener with equal or slight excesses of DLS. The second approach is to blend DLS with sucrose. The most widely evaluated product is a 50/50 blend which is the same as medium invert in sweetness. The blending of DLS and sucrose offers an excellent control over composition because specific amounts of each can be blended. This would eliminate initial product variation caused by the varied levels of residual surcose in commercial medium invert.

TABLE 9.9

COMPARATIVE SWEETNESS OF DEXTROSE-LEVULOSE SYRUP (DLS), SUCROSE (S), MEDIUM INVERT (MI) AND BLENDS IN ACIDIC CARBONATED WATER [1]

	Initial			One Month			Three Months		
	% Solids	% Sucrose[2]	Sweetness Response[3]	% Solids	% Sucrose[2]	Sweetness Response[3]	% Solids	% Sucrose[2]	Sweetness Response[3]
S	10.2	91	N.S.	10.3	55	*Sweeter	10.5	8	*Sweeter
DLS	10.8	0		10.8	0	—	10.8	0	—
MI	10.3	47	*Sweeter	10.4	26	N.S.	10.4	3	N.S.
DLS	10.8	0	—	10.8	0		10.8	0	
MI	10.3	47	N.S.	10.4	26	N.S.	10.4	3	N.S.
50% DLS/50% S	10.3	38		10.4	22		10.4	2	

[1] Prepared in Q-Plus Unit.
[2] % of total sweetener solids.
[3] 20 panelists.
N.S. Not significant.
* p = .05

One important factor has been noted from all tests—no off-flavors have arisen from the DLS. This may result from the fact that the syrup is made from crystalline dextrose.

The results from a 3-month test on cola are shown in Table 9.10. These drinks were bottled by a commercial soft-drink company and evaluated by the same CPC taste panel that evaluated the model systems solutions. Medium invert was the control. No significant differences were noted between the control and the 50% DLS/50% sucrose blend. Also no differences were noted in the DLS blends that contained only 25% sucrose initially. This correlates well to the model system results.

TABLE 9.10

COMPARATIVE SWEETNESS OF DEXTROSE-LEVULOSE SYRUP (DLS) AND DLS/SUCROSE (S) BLENDS AND MEDIUM INVERT (MI) IN COMMERCIALLY FORMULATED COLA SOFT DRINKS[1]

Sweetener	% Initial Solids	1 Week Sweetness Response	6 Weeks Sweetness Response	12 Weeks Sweetness Response
DLS	11.3	N.S.	N.S.	N.S.
MI	11.0			
75% DLS/25% S	11.0	N.S.	N.S.	N.S.
MI	10.9			
50% DLS/50% S	11.1	N.S.	N.S.	N.S.
MI	11.1			
50% DLS/50% MI	11.1	N.S.	N.S.	N.S.
MI	11.0			

[1] 20 panelists
N.S. Not significant

There was no difference between the DLS and medium invert over the entire test period. From the model systems, this would be expected at the 6 and 12-week period after most of the sucrose has hydrolyzed. However, it was surprising that initially the medium invert did not score sweeter. In subsequent tests some panels have detected an initial sweetener difference in favor of the medium invert. These conflicts with the model system results could be caused by some flavoring component which could mask sweetness or by a low sucrose level in the starting commercial medium invert, or by further sucrose inversion before initial taste panel evaluations.

SUMMARY

Dextrose-Levulose Syrup (DLS) and DLS/sucrose blends were compared to sucrose and medium invert in distilled water, acidic carbonated water and formulated soft drinks.

Sucrose was sweeter than DLS at 10% solids, but at 11.5% and above they had similar sweetness. Medium invert, a blend of sucrose-dextrose-levulose, was sweeter than sucrose or DLS. At 25% solids and above, DLS was as sweet as medium invert.

In an acidic environment such as soft drinks, sucrose inverts. Maximum sweetness was observed when 25 to 75% of the sucrose had been inverted. In contrast, DLS gives a constant sweetness because it exists as monosaccharides.

DLS is being used in soft drinks. Two approaches are employed. The first is the use of syrup alone at equal or slight overages to the existing sweetener. The second is to blend with sucrose. The suggested blend is 50 to 75% DLS and 25 to 50% sucrose. Tests in the model systems and soft drinks showed this range of blends to be as sweet as medium invert. A blend of DLS and sucrose offers an excellent control over composition because specific amounts of each can be added. Commercial medium invert does vary widely in residual sucrose. No off-flavors from the DLS were observed in any model or soft-drink formulations.

BIBLIOGRAPHY

AMERINE, M. A., PANGBORN, R. M., and ROESSLER, E. B. 1965. Principles of Sensory Evaluation of Food. Academic Press, New York.

BEEBE-CENTER, J. G., ROGERS, M. S. ATKINSON, W. H., and O'CONNEL, D. N. 1959. Sweetness and saltiness of compound solutions of sucrose and NaCl as a function of concentration of solutes. J. Exp. Psychol. *57*, 231-234.

BERG, H. W., FILIPELLO, F., HINEREINER, E., and WEBB, A. D. 1955. Evaluation of threshold and minimum difference concentrations for various constituents of wines. II Sweetness: the effect of ethyl alcohol, organic acids and tannin. Food Technol. *9*, 138-140.

BIESTER, A., WOOD, M. W., and WAHLIN, C. S. 1925. Carbohydrate studies: I. The relative sweetness of pure sugars. Am. J. Physiol. *73*, 387-396.

BOHM, E., and HOYNAK, P. 1967. Q-Plus: tool for soft drink research. Soft Drinks *85*, No. 1118, 28-30.

BYER, A. J., and ABRAMS, D. 1955. A comparison of the triangular and two-sample taste-test methods. Food Technol. *7*, 185-187.

CAMERON, A. T. 1943. The relative sweetness of sucrose, glucose, and fructose. Trans. Roy. Soc. Can. III. Chem. Math. Phys. Sci. *37*, 11-27.

CAMERON, A. T. 1944. The relative sweetness of certain sugars, mixtures of sugars, and glycerol. Can. J. Res. *22E*, 45-63.

CAMERON, A. T. 1945. The relative sweetness of various sweet compounds and of their mixtures. Can. J. Res. *23E*, 139-166.

CAMERON, A. T. 1947. The taste sense and the relative sweetness of sugars and other sweet substances. Sugar Res. Found., Sci. Rep. Ser. 9.

DAHLBERG, A. C., and PENCZEK, E. S. 1941. The relative sweetness of sugars as affected by concentration. N.Y. State Agr. Exp. Sta. Tech. Bull. *258*, Ithaca, N.Y.

DEER, N. 1922. Relative sweetness of sucrose, levulose, and dextrose. Intl. Sugar J. *24*, 481.

EICKELBERG, E. W. 1940. How to use dextrose in canning. Part 2. Food Ind. *12*, 50-51.

FABIAN, F. W., and BLUM, H. B. 1943. Relative taste potency of some basic food constituents and their competitive and compensatory action. Food Res. *8*, 179-193.

KAMEN, J. M., PILGRIM, F. J., GUTMAN, N. J., and KROLL, B. J. 1960. Interactions of suprathreshold taste stimuli. Quartermaster Food and Container Institute for the Armed Forces. Report 14-60.

KOOI, E. R., and SMITH, R. J. 1972. Newest natural sweetener: dextrose-levulose syrup from dextrose. Food Technol. *26*, No. 9, 57-59.

LAWRENCE, A. R., and FERGUSON, L. N. 1959. Exploratory physiochemical studies on the sense of taste. Nature *183*, 1469-1471.

LICHTENSTEIN, P. E. 1948. The relative sweetness of sugars: sucrose and dextrose. J. Exp. Psychol. *38*, 578-586.

MACKEY, A. O., and JONES, P. 1954. Selection of members of a food tasting panel: discernment of primary tastes in water solution compared with judging ability for foods. Food Technol. *8*, 527-530.

MACKEY, A. O., and VALASSI, K. 1956. The discernment of primary tastes in the presence of different food textures. Food Technol. *10*, 238-240.

MACKEY, A. 1958. Discernment of taste substances as affected by solvent medium. Food Res. *23*, 580-583.

MOSKOWITZ, H. R. 1970. Ratio scales of sugar sweetness. Percept. and Psychophys. *7*, 315-320.

NIEMAN, C. 1960. Sweetness of glucose, dextrose, and sucrose. Mfg. Confect. *40*, No. 8, 19-24, 43-46.

PANGBORN, R. M. 1960. Taste interrelationships. Food Res. *25*, 246-256.

PANGBORN, R. M. 1961. Taste interrelationships II. Suprathreshold solutions of sucrose and citric acid. J. Food Sci. *26*, 648-655.

PANGBORN, R. M. 1962. Taste interrelationships III. Suprathreshold solutions of sucrose and sodium chloride. J. Food Sci. *27*, 495-500.

PANGBORN, R. M. 1963. Relative taste intensities of selected sugars and organic acids. J. Food Sci. *28*, 726-733.

PANGBORN, R. M., and GEE, C. S. 1961. Relative Sweetness of α- and β- forms of selected sugars. Nature *191*, 810-811.

REDFERN, S., and HICKENBOTTOM, J. W. 1972. Levulose-containing corn syrups for the baker. Baker's Dig. *46*, No. 2, 26-27, 30-31.

RENNER, H. D. 1939. Some distinctions in the sweetness of sugars. Confect. Prod. *5*, 255-256.

SALE, J. W., and SKINNER, W. W. 1922. Relative sweetness of invert sugar. J. Ind. Eng. Chem. *14*, 522-525.

SCHUTZ, H. G., and PILGRIM, F. J. 1957. Sweetness of various compounds and its measurements. Food Res. *22*, 206-213.

SPENCER, H. W. 1971. Taste panels and the measurement of sweetness. *In* Sweetness and Sweeteners, G. G. Birch, L. F. Green, and C. B. Coulson (Editors). Applied Science Publishers Ltd. London.

STONE, H., and OLIVER, S. 1966. Effect of viscosity on the detection of relative sweetness intensity of sucrose solutions. J. Food Sci. *31*, 129-134.

STONE, H., OLIVER, S., and KLOEHN, J. 1968. Temperatures and pH effects on the relative sweetness of suprathreshold mixtures of dextrose and fructose. Personal Commun.

STONE, H., and OLIVER, S. M. 1969. Measurement of the relative sweetness of selected sweeteners and sweetener mixtures. J. Food Sci. *34*, 215-222.

TSUZUKI, Y., and YAMAZAKI, J. 1953. On the sweetness of fructose and some other sugars, especially its variation with temperature. Biochem. Z. *323*, 525-531.

VAISEY, M., BRUNON, R., and COOPER, J. 1969. Some sensory effects of hydrocolloid sols on sweetness. J. Food Sci. *34*, 397-400.

WARDRIP, E. K. 1971. High fructose corn syrup. Food Technol. *25*, 501-504.

WHYMPER, R. 1955. The relative sweetness of some sugar solutions—a new angle. Mfg. Confect. *35*, No. 6, 15-16, 19-20, 22.

WILLAMAN, J. J., WAHLIN, C. S., and BIESTER, A. 1925. Carbohydrate studies II. The relative sweetness of invert sugars. Am. J. Physiol. *73*, 397-400.

YAMAGUCHI, S., YOSHIKAWA, R., IKEDA, S., and NINOMIYA, T. 1970A. Studies on the taste of some sweet substances, Part I. Measurement of the relative sweetness. Agr. Biol. Chem. *34*, 181-186.

YAMAGUCHI, S., YOSHIKAWA, R., IKEDA, S., and NINOMIYA, T. 1970B. Studies on the taste of some sweet substances, Part II. Interrelationships among them. Agr. Biol. Chem. *34*, 187-197.

CHAPTER 10

Jonathan W. White, Jr.
J. Clyde Underwood

Maple Syrup and Honey

Maple syrup and honey are sweeteners with much in common. Each has origins obscured in history, each has served as a principal sweetener in the past, and the increasing cost of each has greatly reduced their importance as basic sweeteners in favor of their role as flavorants.

Otherwise, they differ in nearly every respect. Maple syrup is man-made from maple sap, derives its flavor from heat-induced reactions, is essentially a flavored, saturated sucrose solution, and is produced only in the U.S. and Canada. Honey, of course, is a natural product, produced by the honeybee from the nectar of flowers, is ready to be consumed as produced, is essentially a fructose solution supersaturated with glucose, and is a world commodity.

MAPLE SYRUP

When the white man settled in the northeastern section of what is now the U.S. and nearby Canada, he found the American Indian making maple syrup and sugar. The discovery of the secret of the sugar maple tree is shrouded in history, but many nostalgic tales describe the event (Nearing 1950). The Indians had various names for it: sinzibuckwud (drawn from the wood—Algonquin); sheeshee-gummawis (sap flows fast—Ojibway) and sisibaskwat (Cree). Whatever its origin, many settlers quickly recognized the value of maple sugar and learned to make it themselves. However, the product did not become a significant part of the diet in the colonies until the beginning of the eighteenth century when tea and coffee became social beverages (Fox 1905; Raymond 1969). This development brought with it the use of sugar as a sweetener. Since cane sugar was expensive and often very difficult to obtain, the settlers in the Northeast produced and used maple sugar.

The position of West Indies cane sugar in Colonial trade and its taxation greatly increased interest in production of maple sugar. Disruption of sugar trade during and after the Revolution, together with settling of areas west of the original colonies, increased maple sugar production during the 19th century to a maximum of 6 to 6.5 million gal (syrup equivalent) in 1880-1890. During this time, Canadian production increased to over 3 million gal.

TABLE 10.1

PRODUCTION OF MAPLE SYRUP IN UNITED STATES AND CANADA[1]

Year	United States Thousand Gallons (US)	Canada Thousand Gallons (US)
1850	4,282	—
1860	6,613	—
1870	4,477	2,160
1880	6,368	2,570
1890	6,377	3,136
1900	3,548	2,226
1909	5,859	3,476
1919	4,719	3,014
1929	2,509	2,385
1939	2,501	2,616
1949	1,480	2,608
1959	1,049	3,092
1971	962[2]	—

[1] Raymond and Winch 1969.
[2] USDA 1972.

Since 1900, domestic production of maple sugar and syrup has declined as the supply of other sugars—cane, beet, corn—has increased (Table 10.1). As the market price of these commodities fell far below that of maple sugar, the demand for maple sugar became related to its unique flavor, and consumer preference changed to the syrup, 97 to 98% now being sold in this form. Since 1950 maple syrup production in the U.S. had fluctuated between 1 and 1.5 million gal. The most important maple syrup-producing states in decreasing order are (1968 to 1971) New York, Vermont, Ohio, Pennsylvania, Michigan, Wisconsin, New Hampshire, Massachusetts, Maine, and Minnesota. Canadian production, about 3 million gal (US) by 1890, has remained near this level, and the U.S. imports about half of this. With below-average crops in 1969, 1970, and 1971 due to environmental factors which regulate sap flow, the maple syrup producer now faces the possibility that continued short supply and high prices will force large commercial users to seek substitutes.

Making Maple Syrup and Sugar

According to Federal Standards (USDA 1940) "maple syrup" means syrup made by the evaporation of maple sap or by the solution of "maple concrete" (maple sugar). The latter is a form of maple sugar made by concentrating the maple sap far beyond the saturation concentration of the sugar, to about 92% solids. When this hot liquid cools, the sugar crystallizes into a mass for which the term concrete is well-suited. In 50-lb cakes this was a convenient form to store maple syrup. It is too hard for a confection and difficult to redissolve on a small scale. Very little maple concrete is now produced because improved means of storing the syrup have been developed.

The operations involved in collecting the maple sap (known to many farmer-producers as "sweet water") early in the spring and boiling it down to syrup have changed considerably since 1900. No longer, except as a tourist attraction, do we see producers emptying sap-collecting buckets into tanks on horse-drawn sledges and boiling the sap down in an open flat pan over a roaring wood fire in a makeshift lean-to sugarhouse. Today (Underwood and Willits 1963) the modernized maple syrup producer conducts his operations in a stand of sugar maple trees (sugar bush) that has been tree-managed to give him trees spaced properly for vigorous growth and optimum access for sap collection. The trees are tapped with a power drill to a depth of 3 in. and an antiseptic tablet and a plastic "spile" are inserted. When a tree attains a diameter of 10 in. it can support one taphole. For each additional 5 in. another hole may be drilled. To collect the sap, the tapholes are connected into a plastic tubing system which drains into collection tanks or directly into a storage tank at the evaporation plant (sugar house). Where appropriate, vacuum pumping systems are now being installed in plastic tubing collection systems to aid the flow of sap in the lines and even to increase the yield of sap from the tree.

In an enclosed building (sugar house) with a solid floor the sap is evaporated in open, flat pans by atmospheric boiling to a sugar concentration of 65.5% by weight, the minimum concentration in syrup that meets Federal and State specifications. The characteristic maple flavor and color are developed during the evaporation process, which involves exposure to temperatures above 100°C. If the evaporation is done by freeze-drying or under vacuum, only a colorless, flavorless syrup is produced. In other words, flavor and color are not characteristics of the sap as it emerges from the tree, but are developed by the heat applied during concentration. Processing conditions have been established to optimize the production of the light-colored, most distinctively maple-flavored syrups, which bring the highest prices on the retail market. The flat-pan evaporator is efficient (Strolle 1963) and well suited to the making of maple syrup, especially for the small-volume operation usual in the maple industry. It is now being complemented by modern accessories for better control of processing factors. These include the substitution of oil or high-pressure steam for wood as the heat source to obtain a steady maximum rate of evaporation.

Instruments have been developed or adapted for precise control of the end point of evaporation. Standard density syrup (65.5% sugar by weight) boils at 7°F above the boiling point of water at any barometric pressure. Special thermometers are now used for maple sap evaporators that relate syrup temperatures to sugar concentration. Automatic valves withdraw syrup from the evaporator at any desired point of evaporation (Connelly 1969). The flat-pan evaporator is now provided with a tight-fitting cover with a stack for removing steam from the sugar house. This venting system is essential to sanitary operation since it not only keeps foreign materials out of the boiling sap, but also produces a warm, steam-free sugar house. Sanitary practices can now be followed, as in other food-processing plants.

Most maple syrup is still produced by small, individual farm family enterprises. However, centralized sap processing plants are now developing throughout the maple industry, sap collection and evaporation to syrup becoming two separate operations. Improved sap preservation by ultraviolet irradiation has made this practical. Only a limited capital investment and a minimum of labor are needed to put a "sugar bush" into operation and to harvest sap. The central plant can concentrate on making syrup and other maple products and consequently operate on a larger scale. With more syrup, the central plant can do a better job in marketing. A comprehensive description of the new technology is available (Willits 1965).

Composition of Maple Syrup

Maple sap as it exudes from the tree is sterile, and the solids content (usually 2 to 3%) is essentially sucrose, free of reducing hexoses. About 3% of the solids is a mixture of organic acid salts (calcium malate predominating), traces of nitrogen-containing material, a lignin-like substance, and other carbohydrates. If the sap can be reduced to syrup with minimum change in the sucrose, a light-colored, delicately flavored syrup results (Edson 1912; Hayward and Pederson 1946). However, poor-quality sap and/or faulty processing techniques can produce darker, stronger-flavored syrups. Since the syrups with distinct maple flavor and without acrid "caramel" flavor are lighter in color and command higher prices in the market, color is the principal grading factor. Standards have been set by the USDA for interstate commerce (USDA 1940) and by the various states for local sale. Other characteristics, such as density, flavor, and clarity are also considered. The Federal, New York, and Vermont grade designations are listed in Table 10.2.

Chemically, the various grades of maple syrup do not differ except in content of the reducing sugars glucose and fructose, which

TABLE 10.2

GRADE DESIGNATIONS OF MAPLE SYRUP, AS DETERMINED BY COLOR[1]

Grade Designation	Color	Color Index Range[2]
US Grade AA New York Fancy Vermont Fancy	As light as or lighter than Light Amber[3]	0 — 0.510
US Grade A New York No. 1 Vermont A	Darker than Light Amber but as light as or lighter than Medium Amber[3]	0.510 — 0.897
US Grade B New York No. 2 Vermont B	Darker than Medium Amber and as light as or lighter than Dark Amber[3]	0.897 — 1.455
US Unclassified New York No. 3 Vermont C	Darker than Dark Amber	Over 1.45

[1] Willits 1965.

[2] Color index = $A \dfrac{86.3\%}{1 \text{ cm}} = A_{450} \, (86.3/bc)$

A_{450} is the absorbance at 450 nm with distilled water used as a blank;
b is depth of solution in cm; and
c is gm of solids as sucrose per 100 ml of solution.

[3] The terms Light, Medium, and Dark Amber refer to standard glass color filters in the USDA maple syrup color classifier.

TABLE 10.3

COMPOSITION OF MAPLE SYRUP[1]

Component	Amount %	Component	Amount %
Water	34.0	Soluble ash	0.30-0.81
Sucrose	58.2-65.5	Insoluble ash	0.08-0.67
Hexoses	0.0-7.9	Calcium	0.07
Malic acid	0.093	Silica	0.02
Citric acid	0.010	Manganese	0.005
Succinic acid	0.008	Sodium	0.003
Fumaric acid	0.004		

[1] Willits 1965; Hart and Fisher 1971.

increases as the color in the syrup increases, but does not change the sweetening power of the product significantly. A typical composition of maple syrup is shown in Table 10.3. Maple syrup, like other table syrups, has no special nutritive value other than to provide calories. All of them are used primarily to satisfy man's craving for sweet-tasting foods.

In the stiff competition among the many different types and brands of table syrups marketed, consumer preference is generally wooed by the flavor of the particular syrup, and maple syrup meets this competition very well. The true gourmet prefers pure maple syrup to all other table syrups. Most consumers have probably never tasted pure maple syrup. This makes even more startling the overwhelming preference for maple-flavor syrups by the average consumer. A survey conducted by the Homemakers Guild of America (1961) showed that 55% of the consumers contacted preferred maple syrup, 15% corn syrup, 6.6% cane syrup, and 23.1% had no preference. However, pure maple syrup is expensive and most consumers feel they cannot afford this product for daily use. But, because the competition among the national brand table syrups is so great, many companies use the selling power of the maple flavor by marketing products flavored with low percentages of pure maple syrup and/or artificial flavorants resembling maple.

It is clear that the flavorants are now the most important components of maple syrup. Certainly, all the components of maple syrup contribute to its flavor—the sugar, the organic acid salts, and even the oil, butter, or whatever has been used as an antifoam agent during evaporation. However, there is an unknown number of trace materials in the syrup or sugar that gives it "maple flavor". Because these are present in parts per million, they had defied identification for many years, but now, with the modern techniques of gas chromatography and mass spectrometry, progress is being made in solving the mystery of "maple flavor". The flavorants identified to date can be divided (Filipic *et al.* 1969; Underwood 1971) into two groups according to their probable source. One group, possibly formed from ligneous materials in the sap, contains such compounds as vanillin, syringealdehyde, dihydroconiferyl alcohol, aceto-vanillone, ethylvanillin, and guaiacyl acetone. A second group, most likely formed by caramelization of the carbohydrates in the sap, includes acetol, methylcyclopentenolone (cyclotene), furfural, hydroxymethylfurfural, isomaltol, and alpha-furanone.

Uses of Maple Syrup and Maple Sugar

The most direct use of maple syrup or sugar, aside from pure maple syrup or in blends, is in the production of pure maple confections. There are several products worth mentioning as examples of the general types of products. The maple sugar cake is the easiest to make and is the most popular maple confection. This is a small-crystal soft- or hard-textured product made in many forms. Another popular item is a spread called maple cream or maple butter; properly made, it has the consistency of peanut butter and is delicious on toast or crackers. Unfortunately, this product has an uncertain shelf life due to separation of the solid and liquid phases which has limited this really delightful confection to markets local to the producer. A third product containing only maple syrup is crumb sugar, or stirred sugar. This is a granulated sugar product made by boiling syrup to the correct degree of supersaturation and allowing it to crystallize with stirring. Many special variations of these general types are made by enterprising maple products farmers. Details about these can be obtained from the Experiment Stations in the states where the maple industry is active.

Both maple syrup and sugar are used in many foods because of their flavor contributions. Some users have the opinion that maple syrup or sugar is a flavor enhancer, i.e., the effect of certain original flavorants in foods is increased by the addition of maple syrup. Can this be related to its content of methylcyclopentelone, a material now being promoted as a flavor enhancer? Among the food preparations in which maple syrup or sugar are used as an ingredient are cookies, cakes, cake icings, baked beans, candies, ice cream, baked ham, candied sweet potatoes, baked apples, and fresh grapefruit. Recipes for the use of maple in food preparation are available from the Extension Services of many universities and at sales outlets for maple products.

It has been pointed out that maple sugar, since its development as a sweetener during the Revolution, has lost a competitive economic race to cane and beet sugar. However, the American people like maple flavor, and maple syrup continues to be sought by the gourmet. The industry remains small due to the climatic limitations of hard maple growth, the flow of maple sap, and the weather conditions under which maple sap is collected and made into syrup. But perhaps the present growing interest in the use of natural foods may stimulate an increase in the production of maple products.

HONEY

Honey is the sweet viscous fluid elaborated by bees from nectar obtained from plant nectaries, chiefly floral. After carrying it to the hive, the bee ripens this fluid and stores it in the comb for food. "Ripening" is inversion of nectar sucrose and simultaneous concentration of the nectar to about 82% solids. The U.S. Food and Drug advisory definition for honey states that "Honey is the nectar and saccharine exudation of plants, gathered, modified, and stored in the comb by honeybees (*Apis mellifera* and *A. dorsata*); is levorotatory; contains not more than 25% water, not more than 0.25% ash, and not more than 8% sucrose". Although this definition once served a useful purpose, it is considered today to allow much too high a content of water and sucrose and is too low in ash (White *et al.* 1962).

The general characteristics of honey—its sugar composition, color, and flavor—depend upon the kinds of flowers from which it is made by the honeybee. Honey color may vary greatly, from a nearly colorless fireweed or sweetclover type through yellow, yellow green, gold, ambers, dark browns or red browns to the nearly black buckwheat honey. The variations are almost entirely due to the plant source of the honey, though climate may modify the color somewhat by the darkening action of heat.

The flavor of honey also varies over a wide range. A honey may appear to have only a simple sweetness or may be mild, spicy, fragrant, aromatic, bitter, harsh, medicinal, or very objectionable. This is again almost entirely governed by the floral source. In general, a light-colored honey is expected to be mild in flavor and a darker honey to be of pronounced flavor. The exceptions common to all rules sometimes endow a light honey with very definite specific flavors.

Honey is doubtless the oldest sweet known to man, having been the principal sweetener before the discovery of cane sugar. It is probably the only commercially important sweetening material that requires no processing before consumption. For successful handling in commerce, however, it is commonly heated to destroy yeasts and to delay granulation, and generally is either strained or filtered to remove extraneous materials.

About 200 to 250 million lb are produced annually in the U.S. and 15 to 35 million more are imported, largely from Mexico, Argentina, Canada, and Australia. Recent consumer interest in "organic" and natural foods and increases in Japanese imports, together with short crops, have created shortages and prices have doubled since early 1971.

Composition of Honey

The characteristic physical properties of honey—high viscosity, 'stickiness', great sweetness, relatively high density, tendency to absorb moisture from air, and immunity to some types of spoilage—all arise from the fact that natural honey is a very concentrated solution of several sugars. Because of its unique character and its considerable difference from other sweeteners, chemists have long been interested in its composition, and food technologists sometimes have been frustrated in attempts to include honey in prepared food formulas or products.

Average Composition and Variability.—In an analytical survey of U.S. honey (White *et al.* 1962), considerable effort was made to obtain honey samples from all over the U.S. and to include enough samples of the commercially significant floral types that the results, averaged by floral type, would be useful to the beekeeper and packer and also to the food technologist.

The bulletin includes complete analyses of 490 samples of U.S. floral honey and 14 samples of honeydew honey gathered from 47 of the 50 states and representing 82 "single" floral types and 93 blends of "known" composition. For the more common honey types, many samples were available and averages were calculated for many floral types and plant families. Detailed discussions of the effects of crop year, storage, area of production, granulation, and color on composition are included. Some of the tabular data are included here.

Table 10.4 gives the average and the range of values found for each component (White *et al.* 1962).

TABLE 10.4

AVERAGE COMPOSITION OF US HONEY AND RANGE OF VALUES[1]

Characteristic or constituent		Floral Honey	
		Average values	Range of Values
Color[2]		Dark half of white	Light half of water white to dark
Granulating tendency[3]		Few clumps of crystals, 1/8- to 1/4-inch layer	Liquid to complete hard granulation
Moisture	percent	17.2	13.4 - 22.9
Fructose	"	38.19	27.45 - 44.26
Glucose	"	31.28	22.03 - 40.75
Sucrose	"	1.31	0.25 - 7.57
"Maltose"[4]	"	7.31	2.74 - 15.98
Higher sugars	"	1.50	0.13 - 8.49
Undetermined	"	3.1	0 - 13.2
pH		3.91	3.42 - 6.10
Free acidity[5]		22.03	6.75 - 47.19
Lactone[5]		7.11	0 - 18.76
Total acidity[5]		29.12	8.68 - 59.49
Lactone ÷ free acid		0.335	0 - 0.950
Ash	percent	0.169	0.02 - 1.028
Nitrogen	"	0.041	0 - 0.133
Diastase[6] (270 samples)		20.8	2.1 - 61.2

[1] Based on 490 samples of floral honey.
[2] Expressed in terms of USDA color classes.
[3] Extent of granulation for heated samples after 6 months' undisturbed storage.
[4] Reducing disaccharides as maltose.
[5] Milliequivalent per kilogram.
[6] Grams of starch converted by enzyme in 100 gm honey in 1 hr under assay conditions.

The carbohydrate composition of the more common types of honey is shown in Table 10.5. In all cases fructose predominates, although there are a few types, not represented in the table, that contain more glucose than fructose, such as dandelion and bluecurls. This typical excess of fructose over glucose is one way that honey differs from commercial invert sugar. Although honey has less glucose than fructose, the former is the sugar that crystallizes when honey granulates or "sugars". The sucrose level in honey never reaches zero, even though it may contain an active sucrose-splitting enzyme.

The principal physical characteristics and behavior of honey are due to its sugars, but the minor components, such as flavoring

TABLE 10.5

CARBOHYDRATE COMPOSITION OF COMMERCIALLY IMPORTANT HONEY TYPES

No. of Samples	Floral Type	Glucose	Fructose	Sucrose	"Maltose"	Higher Sugars
		%	%	%	%	%
23	Alfalfa	33.40	39.11	2.64	6.01	0.89
25	Alfalfa-sweetclover	33.57	39.29	2.00	6.30	0.91
5	Aster	31.33	37.55	0.81	8.45	1.04
3	Basswood	31.59	37.88	1.20	6.86	1.44
3	Blackberry	25.94	37.64	1.27	11.33	2.50
5	Buckwheat	29.46	35.30	0.78	7.63	2.27
4	Buckwheat, wild	30.50	39.72	0.79	7.21	0.83
26	"Clover"	32.22	37.84	1.44	6.60	1.39
3	Clover, alsike	30.72	39.18	1.40	7.46	1.55
3	Clover, crimson	30.87	38.21	0.91	8.59	1.63
3	Clover, Hubam	33.42	38.69	0.86	6.23	0.74
10	Cotton	36.74	39.28	1.14	4.87	0.50
3	Fireweed	30.72	39.81	1.28	7.12	2.06
6	Gallberry	30.15	39.85	0.72	7.71	1.22
3	Goldenrod	33.15	39.57	0.51	6.57	0.59
2	Heartsease	32.98	37.23	1.95	5.71	0.63
3	Locust, black	28.00	40.66	1.01	8.42	1.90
3	Mesquite	36.90	40.41	0.95	5.42	0.35
4	Orange, California	32.01	39.08	2.68	6.26	1.23
13	Orange, Florida	31.96	38.91	2.60	7.29	1.40
4	Raspberry	28.54	34.46	0.51	8.68	3.58
3	Sage	28.19	40.39	1.13	7.40	2.38
3	Sourwood	24.61	39.79	0.92	11.79	2.55
4	Star-thistle	31.14	36.91	2.27	6.92	2.74
8	Sweetclover	30.97	37.95	1.41	7.75	1.40
3	Sweetclover, yellow	32.81	39.22	2.93	6.63	0.97
4	Tulip tree	25.85	34.65	0.69	11.57	2.96
5	Tupelo	25.95	43.27	1.21	7.97	1.11
7	Vetch	31.67	38.33	1.34	7.23	1.83
9	Vetch, hairy	30.64	38.20	2.03	7.81	2.08
12	White clover	30.71	38.36	1.03	7.32	1.56

materials, pigments, acids, and minerals, are largely responsible for the differences among individual types.

Honey varies tremendously in color and flavor, depending largely on its floral source. Although many hundreds of kinds of honey are produced in this country, only about 25 to 30 are commercially important and are available in large quantity. These may show considerable variation in composition and properties. Until this survey of US honey was reported, the degree of variation was not known, and this retarded the use of honey by the food industry. Legume honeys—clovers and alfalfa and blends—are the most widely distributed and are available in the largest supply.

Water Content.—The water content of honey may range between 13 and 25%. According to the U.S. Standards for Grades of Extracted Honey (USDA 1951) honey may not contain more than 18.6% moisture to qualify for US grade A (US Fancy) and US grade B (US Choice). Grade C (US Standard) honey may contain up to 20% water; any higher amount places a honey in US grade D (Substandard).

These values represent limits and do not indicate the preferred or proper moisture content for honey. Honey with more than 17% moisture and a sufficient number of yeast spores will ferment. Such honey should be pasteurized, i.e., heated sufficiently to kill such organisms. This is particularly important if the honey is to be "creamed" or granulated, since this process results in a slightly higher moisture level in the liquid part. On the other hand, it is possible for honey to be too low in moisture from some points of view. In the western U.S., honey may have a moisture content as low as 13 to 14%. Such honey is somewhat difficult to handle, though it is most useful in blending with high-moisture honey to reduce its moisture content.

In the 490 samples of honey reported in the Department's Technical Bulletin 1261, the average moisture content was 17.2%. Samples ranged between 13.4 and 22.9%, the standard deviation being 1.46; thus 95.5% of all US honey will fall within the limits of 14.3 and 20.1% moisture.

Sugars.—Honey is first and foremost a carbohydrate. Sugars make up 95 to 99.9% of the solids of honey and their identity has been studied for many years. Honey was long thought to be mainly fructose and glucose, with some sucrose and "dextrins". These were considered to be poorly defined complex sugars of high molecular weight. Research in the U.S., Japan, and Canada has shown at least 11 disaccharides in honey in addition to sucrose: maltose, isomaltose, nigerose, turnanose, maltulose (White and Hoban 1959), leucrose, kojibiose (Watanabe and Aso 1960), neotrehalose, gentiobiose, laminaribiose, and isomaltulose (Siddiqui and Furgala 1967). Most of these sugars probably do not occur in nectar, but arise because of either enzymic action during the ripening of honey or chemical action during its storage. The relative amounts of the sugars vary among individual honey types, but all types seem to have the same minor sugars.

Acids.—The acids of honey, though less than 0.5% of the solids, have a pronounced effect on the flavor. They also may be responsible in part for the excellent stability of honey against micro-organisms. At least 18 organic acids have been reported in honey, with varying degrees of certainty. Gluconic acid is now known to be the acid present in the greatest amount in honey (Stinson *et al.* 1960). Its origin will be discussed later. Other acids reported in honey are formic, acetic, butyric, lactic, oxalic, succinic, tartaric, maleic, pyroglutamic, pyruvic, *a*-ketoglutaric, and glycollic (Stinson *et al.* 1960).

Proteins and Amino Acids.—The nitrogen content of honey is low, on the average 0.04%, though it may range to 0.1%. Recent work has shown that only 40 to 65% of the total nitrogen in honey is protein in nature. Of the 8 to 11 proteins shown by gel electrophoresis in various honeys, 4 are common to all, and appear to originate with the bee (White and Kushnir 1967). Free amino acids, containing the remainder of the nitrogen, occur only in trace amounts, proline, glutamic acid, alanine, phenylalanine, tyrosine, leucine, and isoleucine predominating.

Minerals.—Mineral content varies from 0.02% to slightly over 1% for a floral honey, averaging about 0.17% for the 490 samples

TABLE 10.6

MINERAL CONTENT OF HONEY

Mineral	Light Honey ppm	Dark Honey ppm
Potassium	205	1,676
Chlorine	52	113
Sulfur	58	100
Calcium	49	51
Sodium	18	76
Phosphorus	35	47
Magnesium	19	35
Silica	22	36
Iron	2.4	9.4
Manganese	0.30	4.09
Copper	0.29	0.56

analyzed. Values for important minerals are shown in Table 10.6 (Schuette *et al.* 1932, 1937, 1938, 1939).

Enzymes.—One of the characteristics that distinguishes honey, at least in the unprocessed state, from all other sweetening agents is the presence of enzymes, which conceivably arise from the bee, pollen, nectar, or even yeasts and micro-organisms. Those most prominent are added by the bee in the conversion of nectar to honey. Processing and storage may reduce enzyme activities to low levels.

Invertase, added by the bee, splits sucrose into constituent sugars and produces other, more complex sugars in small percentages during this action. This in part explains the complexity of the minor sugars of honey. Although the work of invertase is completed when honey is ripened, the enzyme remains in the honey and retains its activity for some time unless inactivated by heating. Even so, the sucrose content of honey never reaches zero. Perhaps the final low value for the sucrose content of honey represents an equilibrium between hydrolysis and formation of sucrose.

Another enzyme known to be in honey is diastase (amylase). Since this enzyme digests starch to simpler compounds and starch has not been found in nectar, its function in honey is not clear. Diastase appears to be present in varying amounts in nearly all honey. Of all honey enzymes, it has probably had the greatest attention in the past, because it has been, and still is, used as a measure of honey quality (i.e. absence of damage by heating) by several European countries.

Glucose oxidase has been found in honey. This converts glucose to gluconolactone, which in turn forms gluconic acid, the principal acid in honey. Since this enzyme had previously been found in the pharyngeal gland of the honey bee, it is likely that this is the source. Here again, as with other enzymes, the percentage in different honeys is variable. In addition to gluconolactone, this enzyme forms hydrogen peroxide during its action on glucose. This has been shown to be the basis of the heat-sensitive antibacterial activity of honey (White *et al.* 1963).

Other enzymes have been reported in honey, including inulase and phosphatase. Except for catalase (Schepartz and Subers 1966) these have not been sufficiently confirmed.

Food Value

As a carbohydrate food, honey is a most delectable and enjoyable treat. Its distinctive flavors cannot be found elsewhere. The sugars are largely the easily digestible "simple sugars", similar to those of many fruits. Because of its content of such sugars, it is an excellent source of quick energy. It can be regarded as a good food for both infants and senior citizens.

The enzymes of honey, though used as indicators of heating history and hence table quality of honey in some countries, have no nutritional value and are destroyed in the digestive process. The mineral content of honey is variable, but some darker honeys may have significant quantities of trace minerals. Although some vitamins may be demonstrated in honey, the amounts are far too low to have any meaning in human nutrition.

Granulation of Honey

Since the granulated state is natural for most of the honey produced in this country, processing is required to keep it liquid. Careful application of heat to dissolve "seed" crystals and avoidance of subsequent "seeding" will usually suffice to keep a honey liquid for 6 months. Damage to color and flavor can result from excessive or improperly applied heat. Honey that has granulated can be returned to liquid by careful heating. Heat should be applied indirectly by hot water or air, not by direct flame or high-temperature electrical source. Stirring accelerates dissolution of the crystals. For small containers, temperatures of 140° F for 30 min will usually suffice.

If unheated honey is allowed to granulate naturally, several difficulties may arise. The texture may be fine and smooth or granular and objectionable to the consumer. Furthermore, because the water content of the liquid phase increases on granulation, a granulated honey becomes more susceptible to spoilage by fermentation, caused by yeast normally found in all honeys. Quality damage from poor texture and fermented flavors is usually far greater than any caused by the heat needed to eliminate these problems.

Finely granulated honey may be prepared from a honey of proper moisture content (17.5% in summer, 18% in winter) by several processes. All involve pasteurization to eliminate fermentation, followed by addition at room temperature of 5 to 10% of a finely granulated "starter" of acceptable texture, thorough mixing, and storage at 55 to 60° F in the retail containers for about a week. The texture remains acceptable if storage is below about 80° F.

Deterioration of Honey Quality

Fermentation.—Fermentation of honey is caused by the germination and growth of yeasts normally found in all unheated honey. These yeasts, which occur naturally in honey, differ from ordinary yeasts in that they can grow at much higher sugar concentrations than other yeasts, and are therefore called "osmophilic". Even so, there are upper limits of sugar concentration beyond which they will not grow. Thus, the water content of a honey is one of the factors concerned in spoilage by fermentation. The others are extent of contamination by yeast spores and temperature of storage.

Lochhead (1933) has written that honey with less than 17.1% water will not ferment in a year, irrespective of the yeast count. Between 17.1 and 18% moisture, honey with 1,000 yeast spores or less per gram will be safe for a year. When moisture is between 18.1 and 19%, not more than 10 yeast spores per gm can be present for safe storage. Above 19% water, honey can be expected to ferment even with only one spore per gm of honey—a level so low as to be very rare.

Martin (1958) has studied the mechanism and course of yeast fermentation in honey in his work on the hygroscopicity of honey. He confirmed that when honey absorbs moisture, which occurs when it is stored above 60% relative humidity, the moisture content at first increases mostly at the surface before the water diffuses into the bulk of the honey. In such honey, yeasts grow aerobically at the surface and multiply rapidly, whereas below the surface the growth is slower. Fermenting honey is usually at least partly granulated and is characterized by a foam or froth on the surface. It will foam considerably when heated. An odor as of sweet wine or fermenting fruit may be detected. Gas production may be so vigorous as to cause honey to overflow or burst a container.

Honey that has been slightly fermented can sometimes be reclaimed by heating it to 150°F for a short time. This stops the fermentation and expels some of the off-flavor. Fermentation in honey may be avoided by heating to kill yeasts. Minimal treatments to pasteurize honey are shown in Table 10.7 (Townsend 1939; White *et al.* 1963).

TABLE 10.7

PASTEURIZING CONDITIONS FOR HONEY

Temperature °F	Holding Time Min
128	470
130	170
135	60
140	22
145	7.5
150	2.8
155	1.0
160	0.4

Quality Loss by Heating and Storage.—The other principal types of honey spoilage—damage by overheating and by improper storage—are related. In general, changes that take place quickly during heating also occur over a longer period during storage, the rate depending on the temperature. These include darkening, loss of fresh flavor, and formation of off-flavor (caramelization).

To keep honey in its original condition of high quality and delectable flavor and fragrance is possibly the greatest responsibility of the beekeeper and honey packer. At the same time it is an operation receiving perhaps less attention from the producer than any other and one requiring careful consideration by packers and wholesalers. To do an effective job, one must know the factors that govern honey quality, as well as the effects of various beekeeping and storage practices upon it. The factors are easily determined, but only recently are the facts regarding the effects of processing temperatures and storage on honey quality becoming known.

The objective of all processing of honey is simple—to stabilize it, that is, to keep it free of fermentation and to maintain the desired physical state, be it liquid or finely granulated. Methods for accomplishing these objectives have been fairly well worked out and have been used for many years. Probably improvements can be made. The requirements for stability of honey are more stringent now than in the past, since honey is a world commodity, available in supermarkets the year around. Government price support and loan operations require storage of honey, and market conditions may also require storage at any point in the handling chain, including the producer, packer, wholesaler, and exporter.

Application and control of heat is the primary operation in the processing of honey. If we consider storage to be the application of or exposure to low amounts of heat over long periods, it can be seen that a study of the effects of heat on honey quality can have a wide application.

Any assessment of honey quality must include flavor considerations. The objective measurement of changes in flavor, particularly where they are gradual, is most difficult.

As indicators of the acceptability of honey for table use, Europeans have for many years used the activity of certain enzymes and the content of hydroxymethylfurfural (HMF) in honey. They consider that heating honey sufficiently to destroy or greatly lower its enzyme content reduces its desirability for most uses; HMF level is also a measure of heat exposure. A considerable difference has been noted in reports by various workers on the sensitivity to heat of enzymes, largely diastase and invertase, in honey. It is now known that storage alone is sufficient to reduce enzyme content and produce HMF in honey (White *et al.* 1963). Since some honey types frequently exported to Europe are naturally low in diastase, the response of diastase and invertase to storage and processing is of great importance for exporters.

The Codex Alimentarius standards for honey (Codex Alimentarius Commission 1969) require certain combinations of HMF content and diastase activity, depending upon honey type. Because US authorities and honey producers and packers do not agree with the use of HMF and diastase levels as quality indicators, the Codex honey standards have not been accepted in the U.S.

A study of the effects of heating and storage on honey quality was based on the results with three types of honey stored at six temperatures for two years (White *et al.* 1963). The results were used to obtain predictions of the quality life of honey under any storage conditions.

The damage done to honey by heating and by storage is the same. For lower storage temperatures, a much longer time is required to obtain the same result. It must be remembered that the effects of processing and storage are additive. It is for this reason that proper storage is so important. A few periods of hot weather can offset the benefits of months of cool storage: 10 days at 90°F are equivalent to 100 to 120 days at 70°F. An hour at 145°F in processing will cause changes equivalent to 40 days' storage at 77°F.

For storing honey, conditions must be selected that will minimize fermentation, undesirable granulation, and heat damage. Fermentation is strongly retarded below 50°F and above 100°F. Granulation is accelerated between 55° and 60°F and may be initiated by fluctuation at 50° to 55°F. The best condition for storing unpasteurized honey would seem to be below 50°F, or winter temperatures over much of the United States. Warming above this range in the spring can initiate active fermentation in such honey, which is usually granulated and thus even more susceptible.

Some progressive producers and packers are now using controlled temperature storage for honey, particularly in the warmer regions. Using the data from the storage study noted above, we can definitely state that lowering the storage temperature of honey by 10 to 15°F will reduce the rate of deterioration to about one-third to one-sixth of that at the higher temperature. Such a temperature reduction would reduce HMF production to one-third, enzyme loss to one-fifth to one-sixth, and darkening rate to about one-sixth of the rate at the higher temperature. Loss of flavor and freshness would be expected to be reduced similarly. Thus, honey can at any time of the year be more nearly like honey at its very best—fresh from the comb at extraction time.

Marketing and Use

A large part of the honey sold to consumers in the U.S. is in the liquid form, much less in a finely granulated form known as "honey spread" or finely granulated honey, and even less as comb honey. The consumer appears to be conditioned to buying liquid honey, since sales of the more convenient spread form have never approached those of liquid honey. Comb honey has nearly disappeared, though "chunk comb" or comb immersed in liquid honey is sometimes seen in the market. Bulk honey is usually available in 55-gal drums, and tank-truck handling has been reported.

Most industrial use of honey is in baking and other food manufacturing, and in nonprescription cough syrups and tablets. The high fructose content and desirable flavors generally provide customer satisfaction in its uses. It is used in breakfast cereals, syrups, confections, cured meat products, and fruit juices to add a note of "old-fashioned goodness" sought by many customers. It is an optional ingredient for jams, jellies, and preserves.

Several dry products for food manufacture containing a high proportion of honey are commercially available. A pure dry honey product and one containing up to one-third sucrose have been described, but are not in commercial production (Turkot *et al.* 1960). Commercially available materials include Honi-Bake® (Glabe *et al.* 1963) which contains about 23% gelatinized starch. Dry combinations of milk and honey have been described (Walton *et al.* 1951; Webb and Walton 1952; Torr 1966, 1967) but are not commercially available.

Information is available on the use of honey in the manufacture of breads (Smith and Johnson 1951), cake and sweet doughs (Smith and Johnson 1952), cookies (Smith and Johnson 1953A), fruit cake (Smith and Johnson 1953B), and it is a vehicle in medicinal preparations (Rubin *et al.* 1959).

In conclusion, although these two minor sweetening agents, maple syrup and honey, are available in far less quantity than the major sweeteners, their delectable flavors and the unique nature of their origins keep the discerning consumer and creative food technologist interested in using them whenever possible.

BIBLIOGRAPHY

CODEX ALIMENTARIUS COMMISSION, FAO/WHO. 1969. Recommended European regional standards for honey. CAC/RS 12-1969. 23 pp.
CONNELLY, J. A. 1969. Two automatic syrup drawoff controllers. USDA, Agr. Res. Service, ARS 73-60. Washington, D.C. 12 pp.
EDSON, H. A., JONES, C. H., and CARPENTER, C. W. 1912. Microorganisms of maple sap. Vt. Agr. Expt. Sta. Bull. 167, [321] 610.

FILIPIC, V. J., UNDERWOOD, J. C., and DOOLEY, C. J. 1969. Trace components of the flavor fraction of maple syrup. J. Food Sci. *34*, 105-110.

FOX, W. F., and HUBBARD, W. F. 1905. The maple sugar industry. Bulletin No. 59. USDA, Bureau Forestry. Washington, D.C.

GLABE, E. F., GOLDMAN, P. F., and ANDERSON, P. W. 1963. Honey solids—a new functional sweetener for baking. Bakers' Digest *37*, (5), 49-50, 52-54.

HART, F. L., and FISHER, H. J. 1971. Modern food analysis. Springer-Verlag, New York.

HAYWARD, F. W., and PEDERSON, C. S. 1946. Some factors causing dark-colored maple syrup. N.Y. State Agr. Expt. Sta. Bull. 718, 14 pp. illus.

HOMEMAKERS GUILD OF AMERICA. 1961. A consumer survey on flavored syrups; conducted for Owens-Illinois, Toledo, Ohio.

LOCHHEAD, A. G. 1933. Factors concerned with the fermentation of honey. Zentr. Bakteriol. Parasitenk. II Abt. *88*, 296-302.

MARTIN, E. C. 1958. Some aspects of hygroscopic properties and fermentation of honey. Bee World *39*, 165-178.

NEARING, H., and NEARING, S. 1950. The maple sugar book. The John Day Co., New York.

RAYMOND, LYLE S., JR., and WINCH, FRED E., JR. 1959. Maple syrup production in New York State. N.Y. State Col. and Agr., Cornell U., Ithaca, N.Y. 2nd ed., rev.

RUBIN, N., GENNARO, A. R., SIDERI, C. N., and OSOL, A. 1959. Honey as a vehicle for medicinal preparations. Am. J. Pharm. *131*, 246-254.

SCHEPARTZ, A. I., and SUBERS, M. H. 1966. Catalase in honey. J. Apicult. Res. *5*, 37-43.

SCHUETTE, H. A., and REMY, K. 1932. Degree of pigmentation and its probable relationship to the mineral constituents of honey. J. Am. Chem. Soc. *54*, 2909-2913.

SCHUETTE, H. A., and HUENINK, D. J. 1937. Mineral constituents of honey. II. Phosphorus, calcium, magnesium. Food Res. *2*, 529-538.

SCHUETTE, H. A., and TRILLER, R. E. 1938. Mineral constituents of honey. III. Sulfur and chloride. Food Res. *3*, 543-547.

SCHUETTE, H. A., and WOESSNER, W. W. 1939. Mineral constituents of honey. IV. Sodium and potassium. Food Res. *4*, 349-353.

SIDDIQUI, I. R., and FURGALA, B. 1967. Isolation and characterization of oligosaccharides from honey. I. Disaccharides. J. Apicult. Res. *6*, 139-145.

SMITH, L. B., and JOHNSON, J. A. 1951. The use of honey in bread products. Bakers' Dig. *25* (6), 103-106.

SMITH, L. B., and JOHNSON, J. A. 1952. The use of honey in cake and sweet doughs. Bakers' Dig. *26* (6), 113-118.

SMITH, L. B., and JOHNSON, J. A. 1953A. Honey—its value and use in popular cookies. Bakers' Dig. *27* (2), 28-31.

SMITH, L. B., and JOHNSON, J. A. 1953B. Honey improves fruit cake quality. Bakers' Dig. *27* (3), 52-54.

STINSON, E. E., SUBERS, M. H., PETTY, J., and WHITE, J. W., JR. 1960. The composition of honey. V. Separation and identification of the organic acids. Arch. Biochem. Biophys. *89*, 6-12.

STROLLE, E. O., CORDING, J., JR., and ESKEW, R. K. 1956. An analysis of the open-pan maple syrup evaporator. USDA, Agr. Res. Service, ARS 73-14. Washington, D.C. 14 pp.

TORR, D. 1966, Dried honey-milk product. U.S. Patent 3,244,528 April.

TORR, D. 1967, Dried honey-milk product. U.S. Patent 3,357,839 December.

TOWNSEND, G. F. 1939. Time and temperature in relation to the destruction of sugar-tolerant yeasts in honey. J. Econ. Entomol. *32*, 650-654.

TURKOT, V. A., ESKEW, R. K., and CLAFFEY, J. B. 1960. A continuous process for dehydrating honey. Food Technol. *14*, 387-390.

UNDERWOOD, J. C. 1971. Effect of heat on the flavoring components of maple syrup. J. Food Sci. *36*, 228-230.

UNDERWOOD, J. C., and WILLITS, C. O. 1963. Research modernizes the maple syrup industry. Food Technol. *17*, 1380-1382, 1384-1385.

U.S. DEPT. AGR., AGR. MKTG. SERVICE. 1940. United States standards for grades of table maple syrup. Washington, D.C.

U.S. DEPT. AGR., AGR. MKTG. SERVICE. 1951. United States standards for grades of extracted honey. Washington, D.C. 6 pp.

U.S. DEPT. AGR. 1972. Agricultural Statistics. Washington, D.C. 117.

WALTON, G. P., WHITE, J. W., JR., WEBB, B. H., HUFNAGEL, C. F., and STEVENS, A. H. 1951. Manufacture of concentrated milk and honey products. Food Technol. *5*, 203-207.

WATANABE, T., and ASO, K. 1960. Studies on honey. II. Isolation of kojibiose, nigerose, maltose, and isomaltose from honey. Tohuku J. Agr. Res. *11*, 109-115.

WEBB, B. H., and WALTON, G. P. 1952. Dried honey-milk product. U.S. Patent 2,621,128. December.

WHITE, J. W., JR., and HOBAN, N. 1959. Composition of honey. IV. Identification of the disaccharides. Arch. Biochem. Biophys. *80*, 386-392.

WHITE, J. W., JR., RIETHOF, M. L., SUBERS, M. H., and KUSHNIR, I. 1962. Composition of American honeys. U.S. Dept. Agr., Tech. Bull. 1261. Washington, D.C. 1-124.

WHITE, J. W., JR., KUSHNIR, I., and SUBERS, M. H. 1963. Effect of storage and processing temperatures on honey quality. Food Technol. *18*, 555-558.

WHITE, J. W., JR., SUBERS, M. H., and KUSHNIR, I. 1963. How processing and storage affect honey quality. Gleanings in Bee Cult. *91*, 422-425.

WHITE, J. W., JR., SUBERS, M. H., and SCHEPARTZ, A. I. 1963. The identification of inhibine, the antibacterial factor in honey, as hydrogen peroxide and its origin in a honey glucose-oxidase system. Biochem. Biophys. Acta *73*, 57-70.

WHITE, J. W., JR., and KUSHNIR, I. 1967. Composition of honey. VII. Proteins. J. Apicult. Res. *6*, 163-178.

WILLITS, C. O. 1965. Maple syrup producers manual. USDA, Agr. Handbook No. 134, Rev., Washington, D.C.

CHAPTER 11

Karl M. Beck | Synthetic Sweeteners:
Past, Present, and Future

Saccharin was discovered in 1879, so the history of synthetic sweeteners is about 100 yr old. Since that time many other synthetic sweeteners have been investigated but only a few have been used commercially. Progress in the development of synthetic sweetener uses was particularly active from 1950 to 1969, when both cyclamate and saccharin were available. In 1973, saccharin was the only synthetic sweetener permitted in American food, although both cyclamate and saccharin may be used in several other countries. Some other synthetic sweeteners are being developed as commercial products. Sweet taste is a property that has been observed often enough to engender confidence that new synthetic sweeteners will be found in the future.

An overview of the subject of synthetic sweeteners requires considering three aspects of these chemicals. First, with a variety of sugar sweeteners available, why is there any need for a synthetic sweetener? Secondly, what are the criteria of a good synthetic sweetener—what properties should it have? Thirdly, what is the outlook for new synthetic sweetening agents?

There are four types of need for a synthetic sweetener. Three of them are well-known because they have been served in the past by cyclamate, saccharin, and briefly by dulcin, and presently they represent the uses of saccharin. The fourth is a need that will become exceedingly important in the years ahead.

One need for synthetic sweeteners, of course, is for diabetics. The advent of saccharin commercially in 1900 represented a tremendous development for diabetics, for it permitted them to have sweetness in their diet without the hazard of sugar. From 1955 to 1969 the cyclamate-saccharin mixtures in common use permitted the production of a wide variety of good-tasting foods and beverages that were safe for diabetics to consume. There are about 5 million diabetics in the United States, and diabetes is increasing, as there are more older people in the population, who need a synthetic sweetener that can give them a palatable diet.

A second need is for a flavoring agent for pharmaceuticals. Once it was fashionable for medicine to taste bad, and this idea was abetted by an attitude that illness is a divine curse that requires punishment. Drugs have value and they need not taste bad. Synthetic sweeteners often are superior to sugars at masking bitter or unpleasant tastes of drugs. Furthermore, synthetic sweeteners can obviate many technical problems associated with sugars as pharmaceutical excipients, such as hygroscopicity in powders, bulk in tablets, and cap-lock with syrups (Endicott and Gross 1959; Lynch and Gross 1960).

The third need has to do with dieting. A significant part of the American public is overfed, and too many people show it too prominently. There are both physical and psychological aspects to a health problem like obesity. There is a tendency to take the former more seriously, but that may be only because there are better means to measure and ascribe numbers to it. A psychologist once said that inside every fat man is a thin one trying to get out, and the contribution to obesity of the "night-eating syndrome" has been described many times. Overweight has been cited repeatedly as the nation's leading health problem. Statistics are readily available to show that life expectancy is inversely proportional to the extent of excess weight. Obesity particularly is a factor in heart disease, which is the leading cause of death in the United States.

The Food and Nutrition Board of the National Academy of Sciences, recognizing that people are physically less active today, recommended in 1968 that all adults cut down their daily caloric intake. Recommended calorie levels are based on weight, age and sex, but the maximum intake for the average middle-aged male, for example, was lowered from 3,000 to 2,600 Calories per day, and for the comparable female from 2,200 to 1,850 Calories per day. Synthetic sweeteners are a logical approach to calorie control. Most people become overweight gradually by eating a little too much routinely over a period of time. Replacing 100 gm of sugar a day with a noncaloric sweetener can save 400 Calories a day—the reduction recommended by the Food and Nutrition Board—without sacrificing palatability.

Food is abundant in quantity and variety in the U.S., and dieting is not easy. Insurance companies and medical organizations emphasize the health hazards of overweight. Fashion dictates slimness. Most people realize they will look better and live longer if they are not overweight. Nonnutritive sweeteners provide a means of preparing good-tasting foods and beverages with reduced caloric content for use in calorie-controlled diets. Psychologically, they permit calorie reduction without sacrificing palatability or acceptability. That the public likes low-calorie foods and beverages certainly was evident from the sales of such products in the period 1966 to 1969, when cyclamate or cyclamate-saccharin combinations were being used as sweeteners. At that time 76% of the people in the U.S. used some low-calorie products, and 49% of adult Americans were using low-calorie foods and beverages in their approach to dieting as a means of reducing or maintaining weight (Anon. 1969).

The public should be able to have available good-tasting foods and beverages with reduced calorie content if it wants them, and a synthetic, nonnutritive sweetener makes them possible. Certainly, calorie control with an emphasis on essential nutrients—protein, vitamin, and mineral content—is consistent with current concepts of good nutrition.

The fourth need seems to contradict the third, because it concerns the use of synthetic sweeteners for the underfed people of the world—the problems attending scarcity of food. Earth is a planet of fixed dimensions with a finite amount of arable land. About half its population currently do not have enough to eat; there are 39 million people in the U.S., where food abundance creates problems for so many, who do not have enough to eat. World population is growing, particularly in some of the countries which do not now have the agricultural resources to feed their people. Obliterating starvation and producing food to provide even minimal nutritious diets means finding new food sources, especially new sources of protein.

Some of the approaches being studied are quite familiar to food technologists, such as single-cell protein from petroleum, fish flour, and edible algae. These new proteins won't taste like roast beef. At best they are going to be bland, and thus will require flavoring agents. Synthetic sweeteners, which can be produced economically in unlimited quantities, will become important in contriving diets of the future, in an attempt to find a way for everyone in the world to have enough to eat.

Protein receives the bulk of the attention, as it is the most critical in supply and need. But the world probably cannot continue to produce enough sugar for everyone. Corn and sorghum are cheaper and more abundant sources of sugar than are cane and beets, but they also produce starch which can be converted to glucose rather than sucrose. Pound for pound, glucose has as much food value as sucrose, and many scientists consider glucose nutritionally superior to sucrose. However, it is only about two-thirds as sweet as sucrose. Blending a synthetic sweetener with glucose to increase the sweetness could make available sweeter syrups. With synthetic sweeteners it would be possible even to make "double sweet sugar" to let the sugar supply serve more people.

There are several important criteria for a good synthetic sweetener that might be considered prerequisites for a new sweetening agent if it is to be successful commercially (Birch *et al.* 1971). Safety, of course, is the most important requirement. Several potential sweeteners have failed because of toxicity, such as dulcin, P-4000, and stevioside. Saccharin has been in wide use for over 70 yr with few reports of untoward effects, yet its safety is now being questioned. Cyclamate was banned in the U.S. after 19 yr of safe use by the public, even though toxicologists did not agree that the study that led to its removal from the market really indicated carcinogenicity. The experiences with saccharin and cyclamate certainly demonstrate the monumental task of proving safety for a new sweetening agent.

There is widespread feeling that a far more rational approach to the question of safety is necessary. One of the most widely eaten foods is the potato; the most widely used food chemical is salt; the most widely used drug is aspirin. Yet with the current requirements for safety it has been speculated that one could not get FDA clearance for the potato if it were a new food, for sodium chloride if it were a new food additive, or acetyl salicylate if it were a new drug. Perhaps the current regulations need revision. Scientists can understand that living involves risks, and absolute safety is impossible. There is much discussion of the need to use a risk/benefit concept in evaluation safety. This is a concept the government should adopt to do the most good for the most people.

Taste is another important consideration. People are accustomed to the sweet taste characteristic of sugar. The more a synthetic sweetener tastes like sugar, the better its acceptance will be. Cyclamate-saccharin mixtures had a taste that was accepted by most people. The after-taste of saccharin alone has always been a handicap (Helgren *et al.* 1955). Sweetness-time profiles are also important. People are used to detecting the sweetness of sugar in less than a second and having it remain in the mouth for about 30 sec. A synthetic sweetener with delayed or prolonged flavor probably is going to taste "funny". The sweet taste must be compatible with other flavors in food. In pharmaceuticals the ability to mask bitter or unpleasant tastes is important.

Stability is another important criterion. Since food processing often involves cooking, a synthetic sweetener should be stable to the temperatures used in boiling and baking. Foods must be stored, shipped, and stocked on grocery shelves. A sweetener must be stable under various storage conditions, including freezing. The length of stability required will vary with the food product, but for many uses a shelf-life of at least 2 yr is essential. Cyclamate and saccharin have especially good stability. Lack of stability, particularly in liquids, is a limiting factor for the dipeptide sweeteners.

Another important property is solubility. This has been a limiting factor for some of the dihydrochalcones when they are used in place of sugar. While many foods and beverages contain sugar in concentrations of 5 to 15%, products like jellies and maple syrup are more than 50% sugar. To be broadly applicable, a synthetic sweetener should be soluble enough to be able to duplicate the sweetness of simple syrup, which is about 70% sucrose solution.) 1682

Sweetness is one of four basic taste sensations: sweet, sour, bitter and salty. Sourness is a characteristic of protonic organic acids, such as citric, malic, acetic, and tartaric. Saltiness is pretty well limited to the inorganic salts of the Group I metals, especially the halides of sodium, potassium, and lithium; so chemically one doesn't stray far from sodium chloride and still have anything that has a true salt taste. Bitterness and sweetness are not restricted to such well-defined chemical classes (Amerine *et al.* 1965; Moncrieff 1967).

One can readily think of many things that taste bitter, and quickly recognize that bitterness is not limited to any particular class of chemical. Alkaloids, such as quinine, and the xanthines, which include caffeine, generally are bitter. These are heterocyclic nitrogen compounds; but bioflavonoids, such as naringin, which have no nitrogen atom are bitter too. Antibiotics are quite bitter. Many drugs are bitter, such as barbiturates and aspirin. Antihistamines have a persistent bitterness that is quite hard to mask.

Some anthropologists have speculated that bitterness detection developed as a natural warning of poisonous or dangerous substances, and sweetness detection as an indication of pleasant or palatable substances. Perhaps in life one earns more punishments than rewards, because there certainly are far more bitter-tasting than sweet-tasting chemicals. But it is surprising to note that the structural *variety* of chemicals that taste sweet is about as great as the *variety* that taste bitter.

There is a tendency to think of carbohydrates and the closely related polyols as a chemical class of sweet-tasting chemicals, because most people are so familiar with the taste of sucrose, glucose, fructose, lactose, maltose, sorbitol, and mannitol. Particularly when one observes that the sweetness of glycerin is about equal to that of glucose, one is tempted to attribute sweet taste to molecules with multiple HCOH groups. However, of all the monosaccharides, from trioses to hexoses, and the disaccharides and polysaccharides, only a few are sweet. Most are as tasteless as a piece of cotton shirt, which, after all, is a polysaccharide. There are saccharides which are bitter such as gentiobiose. Mannose is an interesting example of the stereospecificity of taste: α-D-mannose is sweet and its stereoisomer, β-D-mannose is bitter (Steinhardt *et al.* 1962).

Several organic salts are sweet, especially the formate, acetate, and propionate salts of lead and beryllium. Lead acetate, known for centuries as "sugar of lead", is the most famous. Several amino acids

TABLE 11.1

RELATIVE SWEETNESS OF VARIOUS ORGANIC CHEMICALS

Chemical	Formula	Sweetness* (sucrose = 1)
sucrose	$C_{12}H_{22}O_{11}$	1
lactose	$C_{12}H_{22}O_{11}$	0.4
maltose	$C_{12}H_{22}O_{11}$	0.5
galactose	$C_6H_{12}O_6$	0.6
D-glucose	$C_6H_{12}O_6$	0.7
D-fructose	$C_6H_{12}O_6$	1.1
invert sugar		0.7-0.9
D-xylose	$C_5H_{10}O_5$	0.7
sorbitol	$C_6H_{14}O_6$	0.5
mannitol	$C_6H_{14}O_6$	0.7
dulcitol	$C_6H_{14}O_6$	0.4
glycerol	$C_3H_8O_3$	0.8
glycine	H_2NCH_2COOH	0.7
sodium 3-methylcyclopentyl sulfamate	$CH_3C_5H_8NHSO_3Na$	15
p-anisylurea	$CH_3OC_6H_4NHCONH_2$	18
sodium cyclohexylsulfamate (cyclamate)	$C_6H_{11}NHSO_3Na$	30-80
chloroform	$CHCl_3$	40
glycyrrhizin	$C_{29}H_{44}O-(COOH)O-C_6H_8O_5-O-C_6H_9O_6$	50
aspartyl-phenylalanine methyl ester	$HOOCCH_2CH(NH_2)CONHCH(CH_2C_6H_5)COOCH_3$	100-200
5-nitro-2-methoxyaniline	$H_2NC_6H_3(OCH_3)NO_2$	167
5-methylsaccharin	$CH_3C_6H_3CONHSO_2$	200

p-ethoxyphenylurea (dulcin)	$C_2H_5OC_6H_4NHCONH_2$	70-350
6-chlorosaccharin	$ClC_6H_3CONHSO_2$	100-350
n-hexylchloromalonamide	$C_6H_{13}CCl(CONH_2)_2$	300
sodium saccharin	$C_6H_4CONNaSO_2$	200-700
stevioside	$C_{38}H_{60}O_{18}$	300
2-amino-4-nitrotoluene	$CH_3C_6H_3(NH_2)NO_2$	300
naringin dihydrochalcone	$C_{12}H_{21}O_{10}(OH)_2C_6H_2COCH_2CH_2C_6H_4OH$	300
p-nitrosuccinanilide	$O_2NC_6H_4NCOCH_2CH_2CO$	350
1-bromo-5-nitroaniline	$H_2NC_6H_3(Br)NO_2$	700
5-nitro-2-ethoxyaniline	$H_2NC_6H_3(OC_2H_5)NO_2$	950
perillaldehyde anti-aldoxime	$CH_2=C(CH_3)C_6H_8CH=NOH$	2000
neohesperidine dihydro-chalcone	$C_{12}H_{21}O_{10}(OH)_2C_6H_2COCH_2CH_2C_6H_3(OCH_3)OH$	2000
5-nitro-2-propoxyaniline (P-4000)	$H_2NC_6H_3(OC_3H_7)NO_2$	4000

*Many factors affect sweetness, and different methods have been used to determine sweetness ratios. The sweetness of sucrose, the usual standard, will change with age due to inversion. Sweet taste depends upon concentration of the sweetener, temperature, pH, type of medium used, and sensitivity of the taster. The usual test methods are dilution to threshold sweetness in water and duplication of the sweetness of a 5 or 10% sucrose solution, although other techniques have also been employed. Where different sweetness values have been reported, the most commonly accepted ones have been cited in this table.
Source: Beck 1969.

are sweet—glycine, histidine, leucine, tyrosine, phenylalanine, and tryptophan. There is a series of sweet-tasting dipetides which will be discussed subsequently. Stereospecificity in taste is common with amino acids; the *D*-isomer of tryptophan is about 35 times as sweet as sucrose, but the L-isomer is not sweet at all. Several naturally occuring substances are sweet, such as glycyrrhizin, from licorice root; stevioside, from an herb that grows in Paraguay; fractions from miracle fruit; and serendipity berries.

A host of completely synthetic chemicals are known to have a sweet taste (Beck 1969; Inglett 1970; Wicker 1966). There are reports describing about 200 such compounds, and one becomes impressed by two things. First is the disparity of structure (Table 11.1). They include chloroform, nitrobenzene, a series of alkoxy-nitroanilines, alkoxyphenylureas, saccharin, cyclohexylsulfamates, alkenols, malonamides, hydrazides, an iminohydantoin, a benzimida-zole, thiazoles, tetrazoles, and several dihydrochlacones. There is no common structural group. Another intriguing aspect is the order of sweetness. Carbohydrate sweeteners—the sweet-tasting saccharides and related polyols—vary in sweetness but over a narrow range: from about one-third to about one and one-third the sweetness of sucrose. Noncarbohydrate sweeteners vary from glycine, which is about two-thirds the sweetness of sucrose, to some synthetic sweeteners that are 5,000 or 6,000 times as sweet as sucrose. Synthetic sweeteners characteristically have a very high order of sweetness.

Looking at the past, present and future of synthetic sweeteners, one can certainly conclude that in the past there has been no shortage of chemicals known to have a sweet taste. The sweet taste of chloroform was reported in 1831; dulcin was discovered in 1883; and saccharin in 1879. Harking back to the criteria of a sweetener in regard to safety, taste, stability, and compatability one can appreciate the importance of these factors by realizing how few synthetics ever became commercial sweeteners. Dulcin (*p*-ethoxy-phenylurea) was on the market briefly a number of years ago, but was withdrawn because it caused liver damage in rats. Saccharin came on the market in 1900 and has been in use continuously since that time. Cyclamate was marketed from 1950 to 1969 in the US, and it still is used in foods and drugs in a number of countries. Mixtures of cyclamate and saccharin were especially well accepted from taste consideration, and met stability and compatability requirements.

Presently, saccharin is the only synthetic sweetener permitted in American foods. Glycyrrhizin is another nonsugar sweetener but is derived from natural sources. Both have some taste limitations— saccharin because some people get a bitter aftertaste from it, and glycyrrhizin because it has a definite licorice taste. Cyclamate and saccharin both are permitted in foods in several other countries, such as Switzerland, Austria, Germany, Norway, and in the form of table sweeteners in Australia, Belgium, Denmark, France, Holland, Canada, etc.

Considerable work is under way with at least two new synthetic sweeteners and one natural nonsugar sweetener. G. D. Searle is developing aspartyl-phenylalanine methyl ester as a synthetic sweetener, and this will be discussed in a later chapter. A Food Additive Petition has been filed by the G. D. Searle Co., so this dipeptide sweetener may be on the market soon. The USDA has been working on dihydrochalcone sweeteners derived from citrus bioflavonoids, especially neohesperidin dihydrochalcone. These sweeteners are discussed in Chapter 16. Considerable work has been done with miracle fruit as a possible commercial sweetener, and this is described in Chapter 18.

All things considered, cyclamate-saccharin combinations seem to be the best synthetic sweeteners developed so far. A lot of food technology was done to create food and beverage products with excellent acceptability based on these sweeteners. Technically, this is the sweetener to beat. The superior sweetener is yet to be found and developed.

Looking further into the future, better synthetic sweeteners probably will be developed. Chemicals with a sweet taste are certainly not a rarity. A problem is the lack of a good theory to account for sweetness. A theory developed by Shallenberger and Acree (1967) has been used to account for the sweet taste of many synthetic sweeteners, but it has failed when used as a means to predict the taste of new sweetener compounds. One has to rely more on luck than theory to find new sweeteners. It is noteworthy that the sweet taste was discovered accidentally in the case of cyclamate, saccharin, the dipeptides, and the dihydrochalcones.

SUMMARY

There are definite needs for new synthetic sweeteners, and all these needs are going to increase. There are hurdles a new sweetener must surmount. They are high, but not insurmountable. Presently there is a gap—a need for a better synthetic sweetener. With the problem outlined, research scientists have a challenge to provide a solution. The future should see new and better synthetic sweeteners.

BIBLIOGRAPHY

AMERINE, M. A., PANGBORN, R. M., and ROESSLER, E. B. 1965. Principles of Sensory Evaluation of Food. Academic Press, Inc., New York.

ANON. 1969. How to cash in on the low-cal revolution. Progressive Grocer *48*, No. 3, 180, 182, 184, 186.

BECK. K. M. 1969. Sweeteners, nonnutritive. Kirk-Othmer: Encyclopedia of Chemical Technology, 2nd Edition *19*, 593-607. John Wiley and Sons, Inc., New York.

BIRCH, G. G., GREEN, L. F., and COULSON, C. B., 1971. Sweetness and Sweeteners. Applied Science Publishers, Ltd., London.

ENDICOTT, C. J., and GROSS, H. M. 1959. Artificial sweetening of tablets. Drug Cosmetic Ind. *85* 176-178.

HELGREN, F. J., LYNCH, M. F., and KIRCHMEYER, F. J. 1955. A taste panel study of the saccharin "off-taste". J. Am. Pharm. Assoc. Sci. Ed. *44*, 353-355.

INGLETT, G. E. 1970. Natural and synthetic sweeteners. Hort. Sci. *5*, No. 3, 139-141.

LYNCH, M. J., and GROSS, H. M. 1960. Artificial sweetening of liquid pharmaceuticals. Drug Cosmetic Ind. *87*, 324-327.

MONCRIEFF, R. W. 1967. The Chemical Senses, 3rd Edition. CRC Press, Cleveland, Ohio.

SHALLENBERGER, R. S., and ACREE, T. E. 1967. Molecular theory of sweet taste. Nature, *216*, 480-482.

STEINHARDT, R. G., CALVIN, A. D., and DODD, E. A. 1962. Taste-structure correlation with α-D-mannose and β-D-mannose. Science *135*, 367-368.

WICKER, R. J. 1966. Some thoughts on sweetening agents old and new, Chem. Ind., No. 41, 1708-1716.

CHAPTER 12

Ferdinand B. Zienty │ Saccharin

When the safety of saccharin was last reviewed, the Food Protection Committee of the National Academy of Sciences (FPC 1970) recommended: (1) long-term feeding studies in two species of animals according to modern protocols to determine chronic toxicity with particular attention to pathological examination of the kidney and urinary bladder for carcinogenic hazard; (2) metabolism studies in man; and (3) studies of toxicological interaction between saccharin and certain drugs. Long-term chronic toxicity studies conducted in several laboratories during the period 1958 to 1970 as shown in Table 12.1 were available to the Food Protection Committee (FPC). A more thorough investigation was desired.

Accordingly, extensive chronic toxicity studies were undertaken in the laboratories listed in Table 12.2. When these studies are completed saccharin will have been more thoroughly tested for safety than most food additives in use today. As results of this new series of investigations are published, detailed protocols, pathological findings and qualified interpretations will become available.

TABLE 12.1

CHRONIC TOXICITY STUDIES ON SACCHARIN 1958- 1970

Investigator	Species	*Product Form
Dr. B. Lessel (1970) Boots Pure Drug Company, Ltd. Nottingham, England	Rat	S
Dr. O. Garth Fitzhugh *et al.* (1951) Food and Drug Administration Washington, D.C.	Rat	SS
Dr. F.J.C. Roe *et al.* (1970) Chester Beatty Research Institute London, England	Mouse	SS

*S = Saccharin (free acid)
SS = Sodium saccharin

TABLE 12.2

CHRONIC TOXICITY STUDIES ON SACCHARIN 1970-1974

Investigator	Species	*Product Form
Dr. L. Golberg Institute of Experimental Pathology Albany Medical College Albany, New York	Rhesus Monkey Rat	SS SS
Dr. B. Lessel Boots Pure Drug Company Ltd. Nottingham, England	Rat	S
Dr. I. C. Munro Dr. H. Grice Health Protection Branch Ottawa, Canada	Rat	SS
Dr. P. Shubik Eppley Institute for Research on Cancer Omaha, Nebraska	Hamster	SS
Dr. L. Friedman Food and Drug Administration Washington, D.C.	Rat Hamster	SS SS
Prof. D. Schmäl German Cancer Centre Heidelberg, Germany	Rat	SS
Dr. T. Miyaji National Institute of Hygienic Sciences Medical Department Osaka University, Japan	Rat	SS
Dr. J. H. Weisburger National Cancer Institute Washington, D.C.	Mouse Rat Rat	SS SS S
Dr. G. J. van Esch National Institute of Public Health Utrecht, Netherlands	Mouse	SS
Mr. P. Derse Dr. P. Nees W.A.R.F. Institute, Inc.** Madison, Wisconsin	Rat	SS

* S = Saccharin (free acid)
SS = Sodium saccharin

** Supported by the International Sugar Research Foundation

Preliminary test results on the recent series of chronic toxicity studies were discussed in a meeting of principal investigators in April, 1972. A recommendation was made that saccharin from different sources be chemically identified and characterized for presence of minor impurities. Analytical work is in progress on lots of saccharin used in several of the chronic toxicity studies to supply the needed information to the principal investigators and to the FPC. Emission spectroscopy and atomic absorption are effective for determination of metals. Thin-layer, gas and high-pressure liquid chromatography are being used for separation of organic impurities, and mass spectroscopy for identification.

The impurities suggested over the years as possibly present in saccharin manufactured by oxidation of o-toluenesulfonamide (Fahlberg and Remsen 1879) were listed by King and Wragg (1966) and by Rader et al. (1967). These and several others that might be visualized are: o-sulfamoyl-benzoic acid, p-sulfamoyl-benzoic acid, o-toluenesulfonamide, p-toluenesulfonamide, ammonium o-sulfobenzoate, ammonium p-sulfobenzoate, saccharin-o-toluenesulfonylimide, toluene-2, 4-disulfonamide, and saccharin-4-sulfonamide.

The first four are the most usual impurities seen in small but variable amounts in commercial saccharin. The last three are rarely detected in saccharin if at all.

Metabolic studies in rats (Kennedy et al. 1972) and in rhesus monkeys (Pitkin et al. 1971) confirm earlier findings (FPC 1955) that saccharin fed at moderate levels is excreted unchanged. Studies in other animals and in man are needed to complete the understanding of saccharin metabolism. Also, a dose-response metabolic study on saccharin should be valuable.

Investigators of saccharin safety have suggested the desirability of making epidemiological studies. No such study has been made over the required long period of use by man, presumably because of the difficulty of collecting the needed information.

Saccharin and its salts were removed from the generally recognized as safe (GRAS) list of substances when a new FDA (1972A) regulation was published. Saccharin is the first substance to be given an interim status under a novel category of food additives regulations newly established by FDA (1972B). The regulatory status of saccharin prior to this point has been summarized (Zienty 1971).

The findings in the new series of chronic toxicity studies will be supplied to the FPC in the spring of 1973 for its reassessment of the safety of saccharin for use as a food additive. The analytical data and new information on metabolism will provide background for interpretation of results. When the FPC reports its recommendations to the FDA the food additive status of saccharin and its salts under the current interim regulation will be reaffirmed or modified.

The 1972 human consumption of saccharin in the United States is estimated at 3.7 million lb.

BIBLIOGRAPHY

FAHLBERG, C., and REMSEN, I. 1879. On the oxidation of o-toluene sulfonamide. Chem. Ber. *12*, 469-473.

FITZHUGH, O. G., NELSON, A. A., and FRAWLEY, J. P. 1951. A comparison of the chronic toxicities of synthetic sweetening agents. J. Am. Pharm. Assoc. Sci. Ed. *40*, 583-586.

FDA, 1972A. Saccharin and its salts. Fed. Regist. *37*, February 1, 2437-2438.

FDA. 1972B. Food additives permitted in food for human consumption or in contact with food on an interim basis pending additional study. Fed. Regist. *37*, December 2, 25705.

FPC. 1955. The safety of artificial sweeteners for use in foods. Nat. Acad. Sci., Publ. 386, Washington, D.C.

FPC. 1970. Safety of saccharin for use in foods. Nat. Acad. Sci., Washington, D.C.

KENNEDY, G., FANCHER, O. E., CALANDRA, J. C., and KELLER, R. E. 1972. Metabolic fate of saccharin in the albino rat. Food Cosmet. Toxicol. *10*, 143-149.

KING, R. E., and WRAGG, J. S. 1966. A rapid method for the estimation of impurities in saccharin and sodium saccharin. J. Pharm. Pharmacol. *18*, Suppl., 22S-27S.

LESSEL, B. 1970. Carcinogenic and teratogenic aspects of saccharin. Proc. SOS (Sci. Survival)/70 Int. Congr. Food Sci. Technol. 3rd, 764-770.

PITKIN, R. M., ANDERSEN, D. W., REYNOLDS, W. A., and FILER, L. J. 1971. Saccharin metabolism in *Mucaca mulatta.* Proc. Soc. Exp. Biol. Med. *137*, 803-806.

RADER, C. P., TIHANYI, S. G., and ZIENTY, F. B. 1967. A study of the true taste of saccharin. J. Food Sci. *32*, 357-360.

ROE, F. J. C., LEVY, L. S., and CARTER, R. L. 1970. Feeding studies on sodium cyclamate, saccharin and sucrose for carcinogenic and tumor-promoting activity. Food Cosmet. Toxicol. *8*, 135-145.

ZIENTY, F. B. 1971. The status of saccharin. Chem. Technol. *1*, 448.

CHAPTER 13

Merrill O. Tisdel
Paul O. Nees
Donald L. Harris
Philip H. Derse

Long-Term Feeding of Saccharin in Rats

With its discovery in 1879, the noncaloric sweetener saccharin began its history, which has been regularly punctuated by debate regarding its safety. As early as 1907, this debate reached such proportions that the President of the United States appointed a review board to evaluate the safety of saccharin. World Wars I and II were periods of high consumption, particularly in Europe, but usage decreased in the post-war period when sugar was again available in adequate supply. During peace time, saccharin was generally limited to individuals who, for medical reasons, could not consume sugar. The introduction of the combination of saccharin and calcium cyclamate provided a more satisfactory product and this, coupled with an increasingly diet-conscious public, resulted in a rapid increase in consumption of non-nutritive sweeteners. Because the newer sweetener had less history of usage and was present at ten times the level of saccharin, the debate on safety shifted from saccharin to calcium cyclamate.

In 1969, cyclamates were banned from use as artifical sweeteners. The increased demand for the artificial sweeteners which had developed in the cyclamate era, was to some extent filled by an increased use of saccharin. Saccharin again became the center of debate on the safety of artificial sweeteners. Fitzhugh *et al.* (1951) reported on the safety of sweeteners and, although they did not conclude that saccharin was unsafe, they did report "increased incidence of the uncommon condition of abdominal lymphosarcoma" in rats receiving diets supplemented with 5% saccharin. Allen *et al.* (1957) reported on a technique for the assessment of bladder carcinogenesis which employed the implantation of pellets containing test materials directly into the bladder lumen of test mice. They reported a significant incidence of bladder tumors in mice receiving saccharin implants. This observation was confirmed by Bryan *et al.* (1971).

In the work being reported, the chronic toxicity of saccharin administered in the diet of rats for a maximum of 100 weeks is examined.

METHODS

Offspring (Fla) from an Fo generation of Sprague-Dawley strain rats were fed for a 100-week test period diets similar to those fed the corresponding parent group. Test diets contained 0%, 0.05%, 0.5% and 5% sodium saccharin in a basal diet (Purina Lab Chow). Twenty male and 20 female weanlings per diet group were housed individually in metal, screen-bottom cages, except that during the mating period males and females were housed in double screen-bottom cages, and from parturition through weaning females were housed in plastic shoe-box cages. The animals were maintained in a temperature-humidity controlled room at $74 \pm 2°$ F with 6 air changes per hr. Feed and water were provided *ad libitum* throughout the study. The animals were observed daily for abnormal appearance or behavior.

Individual body weight and feed consumption were recorded weekly for all animals from week 9 and throughout the study, except during the reproduction study when feed consumptions were not recorded.

Five males and 5 females from each diet group were randomly selected at 13, 52, and 78 weeks on test for clinical data collection. Total RBC and WBC, differential WBC count, hemoglobin, hematocrit, and SGOT data were collected.

At 14 weeks on test, a reproduction study was initiated with all females from the chronic study mated one-to-one with males from the corresponding diet group. The following data were collected for each litter: Number of pups born, number of stillbirths, survival and average pup weight on day 4, survival and individual and average pup weights on day 21, survival and pup weights on day 28, sex ratio of surviving pups. After weaning, pups were group-housed by litter through day 28, examined for gross abnormalities, and sacrificed.

All animals which died on test, sacrificed on test because of a moribund condition, or sacrificed at termination, underwent gross post-mortem examination and the following tissues were collected and preserved in fixative for sectioning and histologic examination: brain, thyroid, lung, salivary gland, intestine (3 levels), adrenal, pancreas, liver, uterus or prostate, kidney, heart, bone, stomach, spleen, bone marrow, urinary bladder, gonad, lumph node, and neoplasms. Heart, liver, spleen, kidneys, and gonads from animals sacrificed at termination were weighed and organ weight to body weight ratios calculated.

RESULTS

Depressed body weights seen in the high level (5%) groups at 9 weeks on test are similar to those which had been seen in the Fo

Generation. These body-weight differences are reflective of smaller weights seen in the litters of the high-level group at weaning of the Fla generation. By 13 weeks on test, however, growth curves are similar for all groups, though body weights of the high-level groups remain slightly smaller. During the last 18 months of the study, few or no meaningful differences exist among the groups regarding body weights (Fig. 13.1).

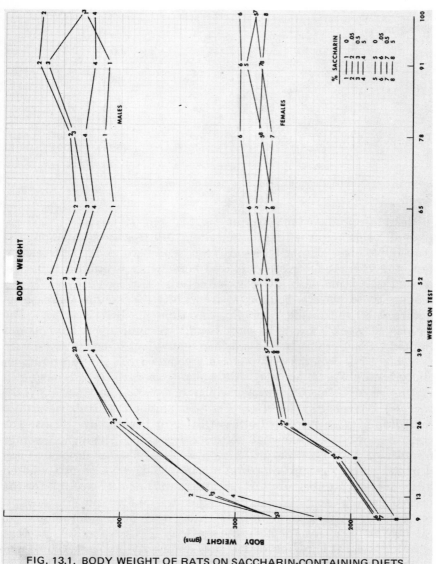

FIG. 13.1. BODY WEIGHT OF RATS ON SACCHARIN-CONTAINING DIETS

No remarkable differences exist among the groups regarding feed consumption throughout the study, though saccharin-fed groups, especially males, tend to have slightly greater average consumptions (Table 13.1).

TABLE 13.1

WEEKLY FOOD CONSUMPTION FOR THE AVERAGE INDIVIDUAL RAT
(gm)

Weeks on Test	Males				Females			
	0%	0.05%	0.5%	5%	0%	0.05%	0.5%	5%
9	139	143	146	141	101	105	102	110
10	142	143	141	150	101	100	100	110
11	133	150	142	149	99	107	104	109
12	131	146	131	144	98	103	101	110
13	129	141	124	145	101	106	105	109
Total	674	723 (723)*	684 (681)	729 (693)	500	521 (521)	512 (509)	548 (521)

* Figures in parentheses indicate amount of basal diet consumed.

Hematology data collected at 13, 52, and 79 weeks on test are generally within normal ranges and no differences are seen among the groups. No gross abnormalities are seen in any group. At 78 weeks on test, the animals show a generalized hemoconcentration due to age, as well as anemia with low platelet counts. Also seen at week 78 is a moderate to severe leukocytosis with an absolute increase in neutrophils and some sporadic elevations in SGOT values. The animals at week 78 were in poor condition clinically but this is not limited to any group and is considered an effect of age (Table 13.2).

Table 13.3 presents group summary data from the reproduction study in the Fla generation. All groups experienced fair success in mating except the 5% group, which had very good mating (90% success). No remarkable differences exist among the saccharin groups regarding average number born per litter or percent live births. The control group had somewhat better performance. The higher average number of pups born per litter and the slightly better survival at day 4 in the control group compared to the 0.5% and 5% groups resulted in an average number of pups per litter surviving day 4 of 9.3 for the control group and 7.1 and 6.9 for the 0.5% and 5% groups, respectively. Average pup weights at day 4 are normal for all groups. Survival is excellent for all groups at day 21 and day 28. Average weaning weights show no differences among the control, 0.05% and

TABLE 13.2

SUMMARY CLINICAL DATA[1]

Group	RBC	WBC	Hgb	Hemat.	Differential (counts/100)					SGOT
					N	L	M	E	B	
				Males 13 weeks						
1	6.93	11.7	16.1	52	16	81	2	1	0	130
2	7.41	8.4	16.4	52	13	86	1	0	0	133
3	7.35	9.6	15.8	51	13	84	1	2	0	122
4	7.68	8.3	15.6	53	20	76	1	2	0	161
				52 weeks						
1	7.04	17.7	15.7	50	25	70	2	3	0	60
2	7.21	20.9	15.3	50	22	71	2	5	0	54
3	7.10	17.8	14.6	47	25	71	1	3	0	56
4	7.33	18.6	14.4	48	25	71	1	3	0	50
				78 weeks						
1	8.35	27.7	14.9	56	41	52	1	6	0	78
2	8.20	31.6	15.2	54	36	60	0	4	0	65
3	8.08	26.7	14.5	51	31	65	0	4	0	78
4	7.71	32.0	14.7	54	46	51	0	3	0	88
				Females 13 weeks						
5	7.96	14.9	16.4	50	13	84	1	2	0	104
6	8.37	12.5	16.3	52	17	79	1	3	0	112
7	7.77	11.4	15.6	48	18	78	0	3	0	79
8	7.47	13.9	15.6	47	14	84	0	2	0	98
				52 weeks						
5	7.03	15.1	15.1	50	14	81	2	3	0	47
6	6.24	17.5	14.0	48	16	80	1	3	0	37
7	6.41	19.1	14.3	47	20	75	2	3	0	39
8	6.61	17.8	13.8	47	17	79	2	2	0	30
				78 weeks						
5	7.72	17.3	13.9	51	32	63	1	4	0	114
6	6.60	21.4	13.1	48	29	66	0	5	0	83
7	7.08	28.3	14.4	49	32	60	2	6	0	96
8	7.52	19.1	14.7	51	25	69	1	5	0	109

[1] Red blood cells (RBC); white blood cells (WBC); hemoglobin (Hgb); hematocrit (Hemat.); neutrophils (N); lymphocyte (L); monocyte (M); eosinophil (E); basophil (B); and serum glutamic oxaloacetic transaminase (SGOT).

0.5% groups, but show markedly lower, below normal average weights in the 5% group. The weight differences persist through 28 days, 7 days after weaning.

Survival data in the chronic study (Fig. 13.2 and 13.3) show no significant differences among the groups through the 100-week test period, though the high-level groups show slightly better survival in females after 12 months on test and in males after 18 months on test.

Organ weight data (Table 13.4) offer few meaningful comparisons among male groups because of the small size of the group at termination. Data from females show few remarkable differences among the groups, especially when considering the large standard deviations and standard errors of the average weights at termination.

Post-mortem examinations revealed gross alterations having the highest incidence to be pneumonia, supporative pneumonia, and animals in generally poor health, along with nephritis in males and

TABLE 13.3

REPRODUCTION STUDY—SUMMARY DATA

| | Percent Saccharin in Diet | | | |
	0	0.05	0.5	5
No. of matings	19	20	19	20
No. of deliveries	15	12	14	18
% deliveries	79	60	74	90
No. born	134*	104	115*	149
No. alive	131	94	98	134
No. dead	3	10	17	15
% live births	98	90	85	90
Av. born/litter	9.6	8.7	8.8	8.3
Av. live births/litter	9.4	7.8	7.5	7.4
Day 4				
No. survivors	130	94	92	124
No. deaths	1	0	6	10
% survival of live births	99	100	94	93
Av. pup wt. (gm)	9.0	9.4	8.5	8.6
Day 21 (weaning)				
No. survivors	98	80	78	113
No. deaths	1	1	0	0
% survival**	99	99	100	100
Av. pup wt. (gm)	45	46	46	38
Day 28				
No. survivors	98	80	78	113
No. deaths	0	0	0	0
% survival***	100	100	100	100
Av. pup wt.	74	73	72	62
Sex ratio M-F	47-53	48-52	45-55	45-55

* One litter eaten—not used in totals or averages
** % survival of survivors day 4
*** % survival of survivors day 21

metritis in females. Other alterations, except for tumors, are nonspecific, sporadic in occurrence within control and test groups, and have a low incidence (Table 13.5).

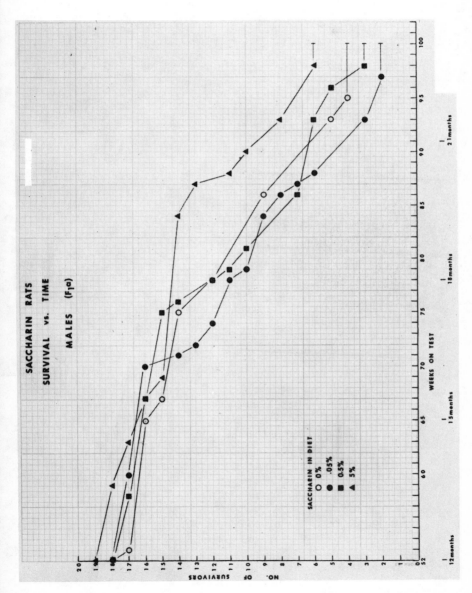

FIG. 13.2. SACCHARIN FED RATS (MALE,F_1A); SURVIVAL VERSUS TIME

Histopathologic alterations believed to be of primary significance to the poor health seen in the animals are chronic and/or supporative pneumonia and tracheitis. The incidence of nephritis is high in the control group females and in all male groups. Other histologic

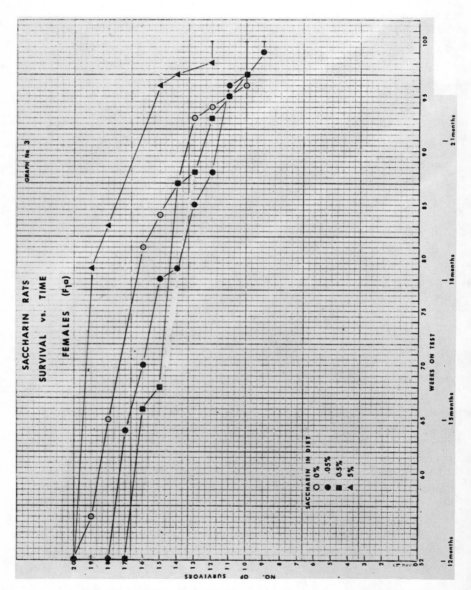

FIG. 13.3. SACCHARIN FED RATS (FEMALE, F_1A); SURVIVAL VERSUS TIME

TABLE 13.4

SUMMARY DATA
ORGAN WEIGHTS — ORGAN WEIGHT RATIOS
TERMINAL SACRIFICE

Group	No. of Animals	Body Wt	Liver	Heart	Spleen	Kidneys R	Kidneys L	Gonads R	Gonads L
			Organ Weights — Males						
1 — 0%	4	420	16.1	1.88	0.955	2.28	2.27	1.81	1.79
2 — 0.05%	2	470	20.8	2.06	1.19	2.93	2.54	1.84	1.75
3 — 0.5%	3	455	19.2	2.14	1.13	2.60	2.44	2.06	1.97
4 — 5%	6	421	15.0	1.80	0.920	2.11	2.14	1.76	1.54
			Organ Weight Ratios — Males						
1 — 0%	4	—	3.84	0.45	0.23	0.55	0.55	0.43	0.43
2 — 0.05%	2	—	4.44	0.44	0.26	0.63	0.55	0.39	0.38
3 — 0.5%	3	—	4.21	0.47	0.25	0.57	0.54	0.46	0.43
4 — 5%	6	—	3.57	0.43	0.22	0.50	0.50	0.42	0.37
			Organ Weights — Females						
5 — 0%	10	275	8.90	1.25	0.572	1.38	1.37	0.0736	0.0939
6 — 0.05%	9	271	9.92	1.34	0.771	1.41	1.41	0.1002	0.0778
7 — 0.5%	10	276	9.73	1.38	0.637	1.44	1.38	0.1119	0.0729
8 — 5%	12	271	9.87	1.30	0.673	1.39	1.38	0.0762	0.0916
			Organ Weight Ratios — Females						
5 — 0%	10	—	3.24	0.45	0.22	0.50	0.50	0.03	0.03
6 — 0.05%	9	—	3.67	0.50	0.29	0.52	0.53	0.04	0.03
7 — 0.5%	10	—	3.55	0.50	0.23	0.52	0.50	0.04	0.03
8 — 5%	12	—	3.63	0.48	0.25	0.52	0.51	0.03	0.03

alterations, except tumors, are nonspecific in nature and of nearly equal incidence in both control and test groups (Table 13.6).

A summary of tumor incidence (Table 13.7) shows a low incidence of tumors in the 0%, 0.05% and 0.5% male groups with a total of 7 tumors in all animals (7/60), but a very high incidence of tumors, a total of 14 tumors in 20 animals, in the 5% male group. All female groups had a high incidence of tumors with totals of 12, 6, 9 and 18 in 20 animals each of the 0%, 0.05%, 0.5% and 5% groups, respectively. Combined tumor incidences of males and females (40 animals/diet) for the 0%, 0.05%, 0.5% and 5% groups respectively are: 15, 8, 11 and 32.

Of all tumors seen, only 6 are seen in 3 or more animals. All others occur in only one or two animals, and vary in incidence among the control and test groups. Of these 6, adenomas of the pituitary are seen in one 0% female, one 0.5% female, and two 5% males. Adenocarcinomas of the pituitary are seen in one 5% male and two 5% females. Adenomas of the thyroid are seen in one 0% female, one

TABLE 13.5

SUMMARY OF MAJOR GROSS ALTERATIONS

	Males				Females			
	0%	0.05%	0.5%	5%	0%	0.05%	0.5%	5%
Total no. animals on test	20	20	20	20	20	20	20	20
No. animals died or sacrificed on test	16	18	17	14	10	11	10	8
No. animals terminal sacrifice	4	2	3	6	10	9	10	12
Alterations observed:								
Animals in poor health	12	11	10	9	7	5	5	5
Lungs: Pneumonia	5	4	9	2	5	1	3	1
Chronic suppurative pneumonia	9	10	8	8	3	5	8	2
Congested and/or hemorrhagic	2	2	3	1	4	2	2	2
Heart: Congested and/or hemorrhagic	2	1	0	0	1	1	0	0
Spleen: Congested and/or hemorrhagic	0	0	0	2	1	0	0	0
Liver: Congested and/or hemorrhagic	4	8	3	4	1	1	2	1
Swollen	0	1	0	0	0	0	1	0
Kidneys: Congested and/or hemorrhagic	4	6	2	4	1	2	0	1
Nephritis/nephrosis	2	5	5	6	0	0	0	1
Testicles: Small/atrophic	2	5	0	1	0	0	0	0
Prostatis or metritis	—	—	—	—	9	7	8	7
P.M. degeneration (no lesion described)	0	0	0	0	0	1	0	0
N.R.A. (no remarkable alterations)	1	0	0	3	4	0	3	2
Tumors: Pituitary								3
Thyroid								6
Ovary					1	1		1
Uterus							1	
Vulva								
Bladder				6				
Sub. cutaneous	2		1	1	6	2	2	5
Abdominal		1				1		
Brain						1		
Total no. tumors	2	1	1	11	11	6	6	17
Total no. animals with tumors	2	1	1	7	8	6	5	13

TABLE 13.6

SUMMARY OF MAJOR HISTOPATHOLOGICAL ALTERATIONS

Alterations: Numbers indicate number of animals/group with this specific lesion.

	Males				Females			
	0%	0.05%	0.5%	5%	0%	0.05%	0.5%	5%
Total no. animals on test	20	20	20	20	20	20	20	20
No. animals died or sacrificed on test	16	18	17	14	10	11	10	8
No. animals terminal sacrificed	4	2	3	6	10	9	10	12
Lungs: Pneumonia	7	2	2	4	8	7	7	9
Chronic suppurative pneumonia	8	11	10	10	5	3	7	2
Congestion and/or hemorrhage			3		1		1	
Pleuritis				1			1	
Tracheitis	9	8	6	11	9	6	11	8
Heart: Congestion	1	1			6	2	1	
Spleen: Congestion	1	3	1	2	2	5	1	0
Splenamegly	1	2	0	2	5	8	1	3
Pigmentation	6	2	3	5	8	9	11	13
Liver: Cloudy swelling	7	4	4	8	9	8	11	13
Vacuolization					1	4	4	4
Pigmentation	6	2	3	1	4	1	1	1
Congestion	6	5	6	4	1	3	3	0
Swollen	3	2	2	1	3	2	3	6
Adrenal: Congestion	3	4	1	2	2	3	1	0
Pigmentation	4	5	0	2	3	4	4	0
Kidney: Congestion	4	4	3	6	4	3	3	2
Pigmentation	0	0	0	1	3	3	3	0
Glomerular atrophy	2	1	0	8	3	0	0	1
Hyaline deposits (tubular)	8	7	7	6	0	0	3	0
Nephritis/nephrosis	5/1	8	7/1	7/2	0	0	1/3	0/2
Oopheritis	—	—	—	—	3	3	3	2
Testicles — degree/metritis (cystic)	9	10	8	9	10	10	12	11
Tumors: See TABLE 13.7								

TABLE 13.7

SUMMARY OF TUMORS—HISTOPATHOLOGY

		Males				Females			
		0%	0.05%	0.5%	5%	0%	0.05%	0.5%	5%
Pituitary	Adenoma				2	1		1	
	Adenocarcinoma				1				2
	Carcinoma				1	1			
Thyroid	Adenoma				1	1	1	1	2
	Adenocarcinoma					1		1	1
Parathyroid	Adenoma								1
Adrenal	Adenoma	1							
Pancreas	Islet cell adenoma					1			
	Islet cell adenocarcinoma								
Ovary	Fibroma							1	1
	Cyst adenocarcinoma								1
	Adenocarcinoma								1
Uterus	Papilloma with sq. metaplasia	1				1			
	Fibroma								
Bladder	Squamous cell carcinoma				1		1	2	1
	Papillary projections						1		
	Undifferentiated malignancy			1*					
	Transitional cell C.A.			1	7				
Subq.	Lipoma	1							
	Fibroma							1	1
	Adenoma							1	2
	Adenofibroma					4	1		
	Fibroadenoma					1	1		
	Fibrosarcoma								
	Sarcoma				1				
	Adenocarcinoma					1			1
	Squamous cell C.A.								1
Abdominal area	Adenoma		1						
Brain	Meningioma		1				1		
Ear	Squamous cell C.A.								1
Totals		3	2	2	14	12	6	9	18

*Epithelial hyperplasia exhibiting numerous mitotic figures and considered precancerous.

0.05% female, one 0.5% female, two 5% females and one 5% male. Squamous cell carcinomas of the uterus are seen only in test females, one 0.05% female, two 0.5% females, and two 5% females. Subcutaneous adenofibromas are seen in four 0% females, one 0.05% female, and two 5% females. Transitional cell carcinomas of the urinary bladder are seen exclusively in seven 5% males. One 0.5% male exhibited epithelial hyperplasia in the urinary bladder that is considered a precancerous type.

Adenomas and adenocarcinomas of the pituitary and adenomas of the thyroid were seen with higher incidence in the test groups than the control groups. The 5 squamous cell carcinomas of the uterus and the 7 transitional call carcinomas of the urinary bladder were seen exclusively in the saccharin-fed groups.

DISCUSSION

Fla generation Sprague-Dawley strain rats held under controlled conditions were fed diets containing 0%, 0.05%, 0.5% and 5% levels of sodium saccharin for a 100-week test period. Growth, hematology, and reproduction parameters were measured. All animals were observed for gross and histologic alterations. Data were reduced and evaluation and comparisons were made among the test and control groups. Under the conditions specified for this study and based on the data presented, certain observations can be made regarding the chronic consumption of saccharin by rats.

The consumption of saccharin had no apparent effect on the body weights of the animals after 13 weeks on test, though body weight depressions had been seen in the 5% groups at 9 weeks on test. There were also no effects on feed consumption or feed efficiency during the period measured.

Though some effects of age and poor health were seen in the hematology data at 78 weeks on test, these effects were seen in all groups, test and control, and cannot be considered a result of consumption of the test material.

Chronic consumption of 0.05% and 0.5% saccharin diets had no remarkable effects on reproductive performance. Data from the 5% group, however, reveal depressed, below-normal pup weights at weaning.

Most major gross and histopathologic alterations seen appear to be nonspecific, sporadic in occurrence within the control and test groups, or have a low incidence, and were apparently not a result of saccharin consumption.

The incidence of tumors in males was apparently increased by the consumption of the high-level (5%) saccharin diet, the incidence being low in the 0%, 0.05% and 0.5% males (3/20, 2/20, 2/20) but very high in the 5% saccharin males (14/20).

Of the tumors which were seen in 3 or more animals, all except the subcutaneous adenofibromas, with 4 in the control (0%) and 3 in the test groups, have a higher incidence in the saccharin-fed groups than in the control groups. These include adenomas and adenocarcinomas of the pituitary, adenomas of the thyroid, squamous cell carcinomas of the uterus, and transitional cell carcinomas of the urinary bladder. The 5 squamous cell carcinomas of the uterus and the 7 transitional cell carcinomas of the urinary bladder were seen exclusively in the saccharin-fed groups, with the higher incidence in the higher levels.

SUMMARY

The artificial sweetener, sodium saccharin, was fed to rats for a maximum of 100 weeks at 0%, 0.05%, 0.5% and 5% of the basal diet. No chronic effects of the consumption are observed in the body weight, feed consumption, or clinical parameters measured. Reproduction data, in general, are normal but markedly depressed weaning weights are seen in the pups of the 5% diet group. A significant increase in tumor incidence is seen in the male 5% diet group. Five squamous cell carcinomas of the uterus are seen in saccharin-fed groups only and 7 transitional cell carcinomas are seen exclusively in males of the 5% diet group.

BIBLIOGRAPHY

ALLEN, M. J., BAGLAND, E., DUKES, C. E., HORNING, E. S., and WATSON, J. G. 1957. Cancer of the urinary bladder induced in mice with metabolites of aromatic amines and tryptophan. Brit. J. Cancer *11*, 212-228.

BRYAN, G. T., and YOSHIDA, O. 1971. Artificial sweeteners as urinary bladder carcinogen. Arch. Environ. Health *23*, 6-12.

FITZHUGH, O. G., NELSON, A. A., and FRAWLEY, J. P. 1951. Comparison of chronic toxicities of synthetic sweetening agents. J. Am. Pharm. Assoc. (Sci. Ed.) *40*, 583-586.

CHAPTER 14

Robert H. Mazur | Aspartic Acid - Based Sweeteners

In December, 1965, James Schlatter was recrystallizing *L*-aspartyl-*L*-phenylalanine methyl ester from ethanol. The mixture bumped and spilled on his hand. Subsequently, when he licked his fingers to pick up a piece of weighing paper, he discovered the remarkable taste of this dipeptide ester. This event catapulted us into an intensive research program on aspartic acid-based sweeteners.

The original objective was the synthesis of gastrin tetrapeptide to be used in a bioassay. Aspartylphenylalanine methyl ester (APM) was an intermediate. Though we did not know it at the time, essentially the same synthesis had been carried out in the pharmaceutical laboratories of Imperial Chemical Industries but they failed to notice the taste of their intermediate.

The taste of APM would not have been predicted from the tastes of the constituent amino acids. L-Aspartic acid is flat while L-phenylalanine is bitter. Two simple amino acids are sweet, glycine and alanine, but these have an intensity (potency) about equal to sucrose. The sweetest amino acids are the D-aromatic ones with potencies in the range 5 to 35. The requirement for D-configuration is rather curious. Potency is the ratio of the sweetness threshold concentration of sucrose in aqueous solution to that of the test compound.

By substituting other amino acids for aspartic acid and for phenylalanine, we discovered that aspartic acid is essential for sweetness but that phenylalanine could be replaced by tyrosine. Based on a value for APM of 150 times sucrose, aspartyltyrosine methyl ester had a potency of 50. The sweetness values given in this chapter were determined by comparison of threshold concentrations of unknowns with sucrose as tasted by 2 or 3 people. In cases where formal taste panel evaluations were carried out, the numbers obtained were similar to our preliminary data. Of the 4 possible diastereoisomers of APM, only the LL was sweet. Also, the β-dipeptide was bitter. Thus, it seemed plausible that a particular receptor site is involved that has a specific geometry with respect to distances and orientation of complexing points. The amide was tasteless. Esters decreased rapidly in potency: Asp-Phe-OC_2H_5, 50; Asp-Phe-OC_3H_7, 1. Our conclusion was that L-aspartic acid is required but that modification of the remainder of the molecule had possibilities (Mazur *et al.* 1969).

We found that the carbomethoxy group could be replaced by methyl and that the requirement for L-configuration still obtained. That is, L-aspartyl-L-amphetamine is a pleasant-tasting compound with a sweetness potency of 50. The benzene ring of this compound could be reduced to cyclohexyl without decreasing sweetness. These changes, in which polar, electron-rich groups were converted to aliphatic and cycloaliphatic substituents, suggested that the size and shape of this part of the molecule is the critical factor in determining sweetness. In other words, the receptor site sees phenylalanine methyl ester as a rather complex amine.

The above results suggested that very simple amides of aspartic acid might be sweet, and this turned out to be true. Some of our findings are shown in Table 13.1. A straight chain of 6 carbons seems to be the smallest alkyl group to give a sweet derivative. In cases where we have resolved the amine, the L-isomer has the desired taste (Mazur *et al.* 1970). Actually, the consistency of the requirement for L-absolute configuration caused us to question and subsequently change literature assignments of absolute configuration of some 1-methylalkylamines (Mazur 1970). An additional discovery was that reasonable potency could be achieved without an asymmetric center in the amine provided that the nitrogen was attached to a fully substituted carbon.

Next we wondered if it would be possible to work some of these analogies backward; that is, would an amino acid with a long

TABLE 14.1

ASP—NHR

RNH_2	Potency[1]
$H_2NCH(CH_3)CH_2CH_2CH_3$	—
$H_2NCH(CH_3)CH_2CH(CH_3)_2$	—
$H_2NCH_2(CH_2)_4CH_3$	1
$H_2NCH(CH_3)CH_2(CH_2)_2CH_3$	30
$H_2NC(CH_3)_2CH_2(CH_2)_2CH_3$	25
$H_2NCH(CH_3)CH_2CH_2CH(CH_3)_2$	50
$H_2NCH_2(CH_2)_5CH_3$	1
$H_2NCH(CH_3)CH_2(CH_2)_3CH_3$	20
$H_2NCH(CH_3)CH_2CH_2CH_2CH(CH_3)_2$	125
$H_2NC(CH_3)_2CH_2CH_2CH_2CH(CH_3)_2$	80
$H_2NCH(CH_3)CH_2(CH_2)_4CH_3$	10
$H_2NCH(CH_3)CH_2(CH_2)_5CH_3$	—

[1]Potency is the ratio of the sweetness threshold concentration of sucrose in aqueous solution to that of the test compound.

aliphatic side-chain give a sweet dipeptide ester? In our original work, all the naturally occurring aliphatic amino acids had given bitter products. Simply straightening out the side chain of leucine to yield aspartylnorleucine methyl ester changed the taste from bitter to sweet. The results with norleucine and higher homologs are shown in Table 14.2. Note that the a-hydrogen can be replaced by methyl without losing sweetness. It would be interesting to resolve these α-methyl-α-amino acids to determine the tastes of the pure diastereoisomeric dipeptide esters.

TABLE 14.2

$ASP-NHC(R_1)(R_2)CO_2CH_3$

R_1	R_2	Potency
CH_3	H(Ala)	—
$CH(CH_3)_2$	H (Val)	—
$CH(CH_3)CH_2CH_3$	H (Ile)	—
$CH_2CH(CH_3)_2$	H (Leu)	—
$CH_2CH_2CH_2CH_3$	H (Nle)	40
$CH_2CH_2CH(CH_3)_2$	H	80
$CH_2CH_2CH(CH_3)_2$	CH_3	200
$CH_2(CH_2)_3CH_3$	H	50
$CH_2(CH_2)_3CH_3$	CH_3	70
$CH_2(CH_2)_4CH_3$	H	70

Returning to the aliphatic amines, by starting with an L-amine and placing a carboxyl alpha to the asymmetric carbon in the long aliphatic chain instead of a couple of methylene groups, an a-amino acid ester is obtained. Examination of a model shows that this new amino acid has the D-configuration. We were thus led to speculate that esters of D-alanine might give sweet aspartyl dipeptides. Remember that LD-APM was bitter. In fact, L-Asp-D-Ala-OMe was sweet, with a potency of 25X sucrose. As the size of the ester increased, the potency increased, peaking at 3 carbons. The data are shown in Table 14.3. This is a reasonable result if the ester is considered as approximately equivalent to the long chain of a 1-methylalkylamine. If I now had a new sweet aspartyl dipeptide methyl ester containing a racemic amino acid, I would make the ethyl ester and predict with considerable confidence that if the potency decreased the LL-isomer had the taste, while if the potency increased, the LD-isomer was the sweet diastereoisomer.

TABLE 14.3

ASP—ALA—OR

R	Potency
CH_3	25
CH_2CH_3	80
$CH_2CH_2CH_3$	170
$CH(CH_3)_2$	125
$CH_2CH_2CH_2CH_3$	10
$CH_2(CH_2)_3CH_3$	6

Subsequent work in this series was carried out with isopropyl esters. These are more stable than methyl esters and hopefully might lead to a product suitable for wet system applications. D-Amino acids with methyl, ethyl, and isopropyl side chains were the best starting materials. Table 14.4 gives these results. Of particular interest is a-aminoisobutyric acid, since this amino acid has no asymmetric carbon and is easy to synthesize on a large scale.

TABLE 14.4

ASP—$NHC(R_1)(R_2)CO_2CH(CH_3)_2$

R_1	R_2	Potency
CH_3	H (Ala)	125
CH_3	CH_3 (Aib)	50
CH_2CH_3	H (Abu)	170
CH_2CH_3	CH_3 (Iva)	50
CH_2CH_3	CH_2CH_3	12
$CH(CH_3)_2$	H (Val)	170
$CH(CH_3)_2$	CH_3	20
$CH_2CH_2CH_3$	H (Nva)	17
$CH(CH_3)CH_2CH_3$	H (Ile)	4
$CH_2CH(CH_3)_2$	H (Leu)	—

This is a fundamental, novel, and unpredictable discovery of a general structure that will trigger a sweet response. Generally, if you have L-isoasparagine with the amide nitrogen attached to a carbon bearing two groups of sufficiently dissimilar size and having the proper absolute configuration, then the compound will be sweet.

ACKNOWLEDGMENT

For their chemical insight and experimental skill, my sincere appreciation is expressed to James Schlatter, Arthur Goldkamp, Patricia James, Judith Reuter, and Kenneth Swiatek.

BIBLIOGRAPHY

MAZUR, R. H., SCHLATTER, J. M., and GOLDKAMP, A. H. 1969. Structure-taste relationships of some dipeptides. J. Am. Chem. Soc. *91*, 2684-2691.

MAZUR, R. H., GOLDKAMP, A. H., JAMES, P. A., and SCHLATTER, J. M. 1970. Structure-taste relationships of aspartic acid amides. J. Med. Chem. *13*, 1217-1221.

MAZUR, R. H. 1970. Absolute configuration of 1-methylalkylamines. J. Org. Chem. *35*, 2050-2051.

CHAPTER 15

Charles I. Beck

Sweetness, Character, and Applications of Aspartic Acid-Based Sweeteners

Selection of sweetener candidates for further development is based upon a weighted consideration of a number of factors: safety (toxicity, metabolic rate, etc.); taste (character and intensity); physical-chemical characteristics (sufficient solubility to perform as a sweetener; stability in target applications alone or in presence of other food chemicals; stability as bulk chemical; lack of hygroscopicity; appearance); cost (to achieve the necessary sweetness); and legal (both from the point of view that it is lawful to sell based on other present art and from the position of patent protection).

The steps used in the taste evaluation of sweeteners are given in Table 15.1.

TABLE 15.1

STEPS IN TASTE EVALUATION OF SWEETENERS

Safe for tasting
Preliminary expert evaluation
Sweetness character
Sweetness intensity
Evaluation in target applications

Safety

Safety evaluation of potential sweeteners, as with any other potential food ingredient, should be run at least to the extent required for the test intended. Safety here refers to toxicity tests and other appropriate screening tests for biological activity. Thus, a test in which the product is tasted and expectorated might not require as much safety testing as a product that would be consumed.

Expert Evaluation

Expert evaluation is carried out by one person, or a few people, familiar with the performance requirements and specifically with food technological and organoleptic criteria. This evaluation serves two purposes: (1) it refutes or confirms suspicions, allegations, and/or sneak previews; and (2) it provides a basis for designing tests, for both expert and naive panels. If the particular product is not worthy of further testing, the tests stop here. If it shows any promise, there are enough data to set initial testing conditions and to draft a relevant questionnaire.

The next two steps—determination of taste character and sweetness intensity—present problems because (1) it is difficult to establish sweetness intensity if one is not comparing the same taste quality; and (2) it is difficult to establish taste quality if one is not at the same sweetness intensity (e.g., the bitterness of saccharin increases substantially as the concentration or sweetness level of the application increases). Usually, one or two tests are adequate to focus on the proper test conditions.

Sweetness Character

In addition to establishing that the compound imparts a sweet taste, it becomes important to determine whether that sweet taste is in fact perceived to be the same as that of sucrose and other normally accepted nutritive food sweeteners, and whether the sweetness intensity builds and then recedes in the mouth, as a function of time, in an acceptable way. If there are other tastes, can they be tolerated? If there are other intensity profiles, will they be acceptable? Definitive work in flavor characterization is most commonly done by the flavor profiling technique reported by Caul (1957).

Sweetness Intensity

Measuring the sweetness intensity of a new compound is, on the surface, a straightforward task. However, if one wishes to take into account the many factors which may alter the result of this determination, one is faced with the reality that there is no single answer. In a water system, most sweeteners appear to show less sweetness intensity as their concentration increases. Saccharin is a classic example of this phenomenon, and Salant (1968) tabulated potency values which ranged from 500 to 200 versus sucrose controls, which ranged from 2 to 20%. Since many sweeteners seem to respond in this way, it is interesting to postulate that perhaps it is the nonlinear nature of the sweetness of the sucrose standard that causes this phenomenon.

In addition to concentration, the nature of the food medium can also have a significant effect on the perceived sweetness intensity of a given compound. Thus, it would not be too surprising if the given sweetener was found to have different potencies in vanilla pudding and Italian-flavored salad dressing (even if the sucrose percentages were the same). Sweetness is a basic taste, but nonetheless it is a flavor, and as such it must be compatible with the total food system.

Even minor differences in character will affect the performance of sweet compounds.

There are many methods of determining sweetness; in fact, after reviewing the methods, it becomes apparent that there are various interpretations of what is meant by sweetness or sweetness intensity. There is an abundance of literature that refers to the threshold determination of sweetness, as well as other basic tastes; and this has been well reviewed by Amerine *et al.* (1965). For sweetness, the threshold is the lowest concentration at which not only the presence of the compound is apparent, but at which its character is recognized as sweetness. Working with thresholds is difficult because they vary with the individual over fairly wide ranges, and most application work takes place at substantially higher concentration levels. An advantage of the threshold method is that only small amounts are required or ingested. Another method which has been extensively applied to sweetness is termed ratio scaling (Stevens 1956; Stone and Oliver 1969; Moskowitz 1970A, 1970B and 1971). In ratio scaling the sensory judgments of sweetness are obtained by magnitude estimation, a method in which numbers are assigned to stimuli in proportion to their perceived sweetness. This methodology has been primarily used in model aqueous systems, though it should be generally useful.

Additional work has been done on comparison of the sweetness of sweeteners and sweetener mixtures. Morton (1972) deals with a population distribution method of sweetness in a given formulation based on effective dose $_{50}$ (ED_{50}) methodology. This method gives potency by plotting population percentage preferring the test sweetener to a standard sucrose control versus the concentration of the test sweetener. Potency is obtained by dividing the standard sucrose control concentration by the test sweetener concentration at which half of the population preferred the test sweetener and half preferred the sucrose control. The answers obtained by this method do give direct readings of potency; however, large panel sizes are required in order to diminish standard deviations. It is also possible to utilize roughly the same test samples being presented to test the hypothesis that the two samples are identical. In this case, the exact identity of the potency is not determined, but the confidence level can be high.

Evaluation in Target Applications

Assuming that the candidate compound has passed the screening tests described, it is now known that it is safe, at least for testing, and potentially effective as a sweetener (it has a pleasant and sufficiently intense sweetness). The first step in evaluation is to determine for what use the sweetener is intended, that is, as an across-the-board substitute for sugar, use in a specialty diabetic product, or use in low-calorie food. As a replacement for sucrose, many food technological problems arise. If all the sucrose is taken out of a candy, cookie, or cake, it becomes necessary to devise other means of filling up that space, or bulking the product. If one is trying to promote low-calorie foods, it becomes necessary to bulk the product to maintain the calorie reduction which the use of an intense sweetener permits. Besides sweetness and bulk, sucrose often provides other important physical properties in food products: protection against microbial activity due to high osmotic pressure, body (mouth-feel), browning (source of glucose), humectancy, etc. Furthermore, sucrose in its applications must withstand not only the temperatures used for bakery products, and the much higher temperatures of candy making. Oddly enough, the performance stability of sucrose is based not on its chemical stability, but on the sweetness of its decomposition products, glucose and fructose. All these factors enter into the decision of what applications should be chosen as the test systems in product evaluation.

For initial screening it may be necessary only to apply the sweetener in those major target applications in order to establish that it does perform in an acceptable way. However, sooner or later it will become necessary to determine that the sweetener can be applied in formulations without: (1) interacting with other ingredients, (2) developing off-taste and (3) significant loss of sweetness. Before the product is fully developed it will be necessary to evaluate it in every possible class of application for which it is intended.

Naturally the traditional test of shelf-life stability in finished formulations and performance in shelf-life stability for each of the product forms (tablet, liquid, spoon-for-spoon, and bulk) will need to be ascertained.

Lastly, now that all the logical uses of the product have been evaluated, it becomes necessary to evaluate its illogical uses. Therefore, it will be necessary to look at oversweetened, overheated, and otherwise unintended applications.

TASTE STUDIES ON
ASPARTYLPHENYLALANINE METHYL ESTER

Aspartylphenylalanine methyl ester has been selected from the various sweet aspartic acid derivatives mentioned earlier for extensive development. In addition to its chemical name it has been referred to as dipeptide sweetener, APM, and aspartame. There are two previous scientific articles which deal with the performance of aspartame (Cloninger and Baldwin 1970; Mazur and Craig 1970).

Sweetness Character

Basically, it appears as though aspartame has a taste which is very similar to sucrose (Mazur *et al.* 1969). There is a slightly longer period of time required for aspartame to reach its peak sweetness compared to that of sucrose. This difference is detectable at first contact, but distinction thereafter becomes tenuous. After-taste does not appear to differ from sucrose in normally sweetened foods. Table 15.2 lists the flavor profile and time sweetness relationship for aspartame and an aspartame:sodium saccharin (5:1) mixture by weight, when compared to sucrose at a sweetness level, in water, equivalent to 10% sucrose.

TABLE 15.2

FLAVOR EFFECTS OF ASPARTAME STUDIED AT A SWEETNESS LEVEL EQUIVA-
LENT TO 10% SUCROSE IN WATER

Sweetener	Concentration %	Flavor Profile Description	Time-Intensity Relationship for Basic Sweet Curve
Sucrose	10	Almost entirely a sweet taste with barely noticeable drying. Has a sensation of full- ness	Immediate sweetness which disappears evenly and rapidly
Aspartame	0.075	Sucrose-like sweet- ness with very little synthetic sweetener flavor effects	Slight delay in sweetness peak; otherwise similar to sucrose
Aspartame: Sodium Saccharin (5:1)	0.033	Sucrose-like sweet with some bitter- ness, but very slightly more synthetic sweetener sensation than aspartame	Slightly delayed sweetness, but otherwise like aspartame and sucrose

In the taste work which has been done so far, all people sampled seem able to taste the sweetness of aspartame.

Sweetness Intensity

The sweetness intensity or potency (weight of sucrose per weight of sweetener giving equal sweetness) of aspartame decreases with sucrose concentration. Table 15.3 shows aqueous concentrations of sucrose and aspartame, and the resulting potency values.

TABLE 15.3

ASPARTAME POTENCY AS A FUNCTION OF SWEETNESS LEVEL (IN A SIMPLE AQUEOUS SYSTEM)

Concentration (%)		
Sucrose	Aspartame	Potency
0.34*	0.007-0.001	400
4.3	0.02	215
10.0	0.075	133
15.0	0.15	100

*recognition threshold

Because of the importance of table top uses of sweeteners, a coffee model was used as a test system. Potencies of various formulations were then determined versus a 4% sucrose control. Fig. 15.1 shows a population distribution of sweetness identification as a function of assumed potencies, and the simple questionnaire used to gather the data. It should be observed that an increase in assumed potency represents a decrease in concentration of aspartame.

Coffee samples were served, randomly coded, to judges at about 60°C. Judges were never asked for more than two judgements at one sitting. The coffee was prepared with 5.5 gm of instant coffee (standardized) per liter (the weakness of the coffee of about half strength, was intended to increase sensitivity of the judges to the sweetness change). If the same panelists are used for each point, it is possible to obtain a standard deviation by using a method of probits (Finney 1947).

The effect of body (mouthfeel) was eliminated by the fact that the judges were instructed to select the sweetest sample, not the sucrose or aspartame one. Moreover, when the panel was tested on its ability to select the sample which was most "watery", they were unable to sense the difference. It is interesting that the potency of aspartame in coffee is not significantly different from that in water, when both systems are evaluated near the level of 4% sucrose. When evaluating sweetness in coffee it is not always clear whether judges are responding to sweetness, or to the masking or removal of coffee bitterness. In strong coffee the apparent potency of aspartame increases.

When the potency of eight lots of aspartame was tested using the graphical procedure just shown, the potency was found to be 209

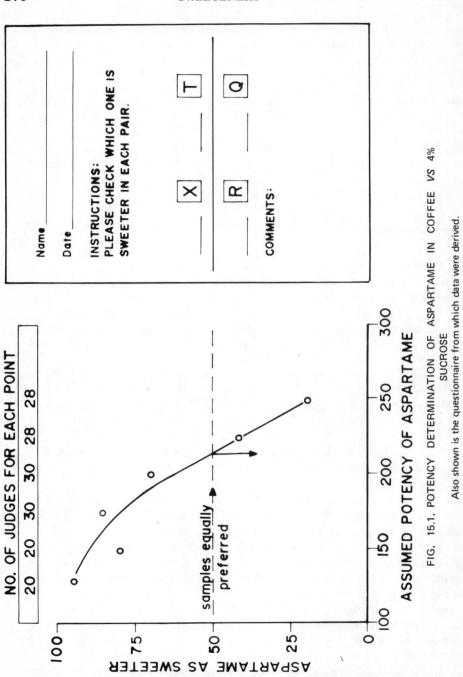

FIG. 15.1. POTENCY DETERMINATION OF ASPARTAME IN COFFEE *VS* 4% SUCROSE

Also shown is the questionnaire from which data were derived.

plus or minus 7, based on pure aspartame. Generally, two well selected points are adequate to obtain an indication of the potency. Figure 15.1 shows our first attempt with this method, and as such provides an indication as to the shape of the curve.

Aspartame has been evaluated in a wide variety of products which were selected to represent the major categories of use. Table 15.4 indicates the range of sweetness potencies observed for aspartame in the various applications. Naturally, the potencies can vary with the sweetness and flavoring of a given formulation, but most seem to fall between 160 and 220.

All the preceding work was done at aspartame levels that were calculated, and then determined by the panelists to be correct and acceptable, compared to known standards. Additional work was done to establish the *criticality* of properly sweetening a formulated food product. In this case, the simple system of coffee was used again, but samples were presented on a single stimulus basis and were evaluated for their acceptability on a sweetness basis. Figure 15.2 shows a dramatic insensitivity on the part of the panelists: thus, the aspartame content of coffee had to be doubled to move the average

TABLE 15.4

POTENCY OF ASPARTAME IN VARIOUS APPLICATIONS

Application	Potency	Sucrose Level in the Comparison Standard (as prepared %)
Drinks		
powdered beverage mixes	180	11
soft drinks	180	10
coffee (spoon-for-spoon)	180	4
tea, hot (spoon-for-spoon)	180	6
tea, iced (spoon-for-spoon)	180	7
cold fruit drinks (spoon-for-spoon)	180	14
Confections		
chocolate or cocoa products	200	58
Breakfast cereals		
presweetened	180	37 (dry)
unsweetened (spoon-for-spoon)	220	—
Desserts		
gelatin	160	18
pudding	200	17
frozen desserts (Ice Milk)	280*	14 (including other nutritive sugars)
Raw fruit (spoon-for-spoon)	220	—

*This high figure may be accounted for by synergy between aspartame and sorbitol in the low calorie ice milk formula. Fructose is also known to give high sweetness in frozen dessert.

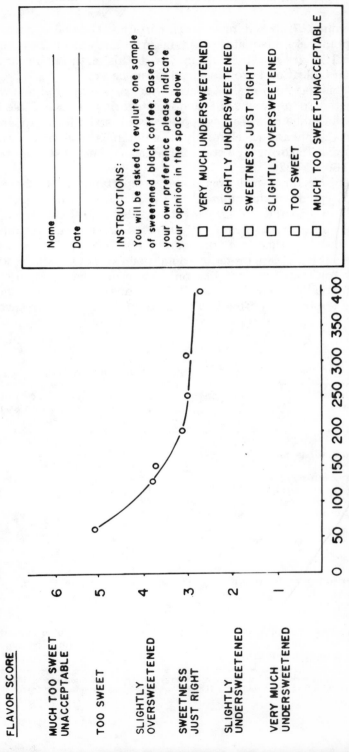

FIG. 15.2. EFFECT OF SWEETENER CONCENTRATION ON SWEETNESS ACCEPTABILITY

Also shown is the questionnaire from which data were derived.

response of the panel from sweetness "just right" to "slightly over-sweetened"; similarly, when the aspartame content was halved the average panel response barely moved from "sweetness just right". At the extremely high levels of sweetness in Figure 15.2 (low assumed potencies) the unacceptability of the coffee was based primarily on its excess sweetness, but in part was also due to a slight chemical taste.

These data have shown that aspartame has a sweetness similar to that of sucrose and an intensity of about 160 to 220 times sweeter than sucrose on a weight basis. However, there was concern with possible flavors, since under some conditions the compound can cyclize to give a conversion product, namely a diketopiperazine according to Davey *et al.* (1966). The rate of this reaction is controlled by: (1) limiting the storage time in aqueous systems; (2) seeking to keep the compound near the pH range of 3 to 5; and (3) minimizing the heat applied to the compound.

Because this reaction could occur in aspartame-sweetened foods, taste effects other than loss of sweetness, were studied, but there were none. The conversion product was blended with aspartame and tasted via two protocols: (1) the conversion product was substituted for up to 50% of the aspartame; (2) the conversion product was added over and above the normal aspartame by as much as half of the aspartame. These tests were carried out in both coffee and lemonade. The only taste effect of aspartame conversion appears to be the loss of sweetness (Figs. 15.3 and 15.4). It can be seen that the 50% points (where the judges are equally divided as to whether the 4% sucrose or the test sample is sweeter) in almost all cases lie within the potency range of 175-200. Thus, the points above the cross hatched area (except for one) are less than 50% finding sucrose sweeter, and all the points below the cross-hatched area are above 50%. No consistent identification of the conversion product occurred.

Taste-modifying substances have potentiating and synergistic effects in the presence of other flavors. There are some indications that aspartame can mellow coffee flavor, intensify fruitiness, etc., but detailed proofs to establish this property are difficult on a datum wherein sweetness is changing.

The synergism of aspartame and saccharin was indicated as follows. Data were developed, in a simple aqueous system, which indicated sweetness equivalence for sucrose solutions with: (1) saccharin solutions and (2) aspartame solutions at various levels of sucrose concentration. In addition, the concentrations of various ratios of aspartame: saccharin necessary to develop sweetness equal to 10% sucrose were determined. A traditional approach to testing

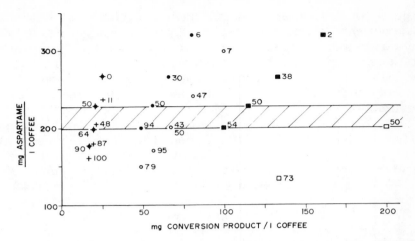

FIG. 15.3. EFFECT OF CONVERSION PRODUCT ON THE SWEETNESS OF ASPARTAME-SWEETENED COFFEE

Numbers adjacent to points are % judges finding the 4% sucrose control sweeter than the test sample. The cross-hatched area represents the anticipated potency range (175-200) of aspartame. +, ○, □, are 10%, 25% and 50% replacement of aspartame with the conversion product; ✦, ●, ■ are 10%, 25% and 50% additions to aspartame.

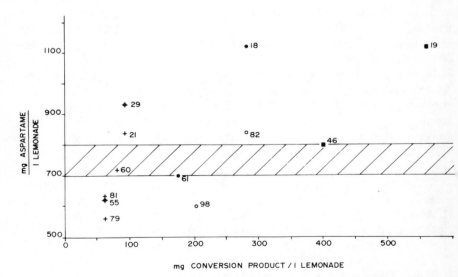

FIG. 15.4. EFFECT OF CONVERSION PRODUCT ON THE SWEETNESS OF ASPARTAME-SWEETENED LEMONADE

Numbers adjacent to points are % of judges finding the 14% sucrose control sweeter than the test sample. The cross-hatched area represents the anticipated potency range (175-200) of aspartame. +, ○, are 10% and 25% replacement of aspartame with the conversion product; ✦, ●, ■ are 10%, 25%, and 50% additions to aspartame.

for synergy would be to add individual calculated contributions to sweetness to see if you would expect less than was observed (10% sucrose equivalence for this case). The following equation could be used:

$$S = C_A \times P_A + C_S \times P_S$$

where: S is the calculated total sweetness of the system in percent sucrose units;

C_A, C_S are concentrations of the aspartame and saccharin, respectively; and

P_A, P_S are potencies of the aspartame and saccharin, respectively, each versus sucrose. Each potency is determined from its concentration.

In Fig. 15.5, the calculated total sweetness of the system is plotted versus aspartame: saccharin ratios, on a weight basis, and is shown as the (+) points. Synergism is indicated by the points that are below the observed sweetness, equivalent to 10% sucrose.

In the foregoing traditional analysis, it is assumed that the potency of any sweetener is only a function of the concentration of that

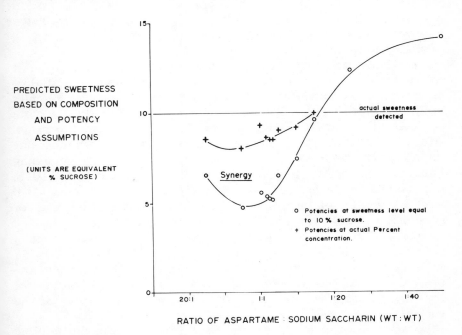

PREDICTED SWEETNESS BASED ON COMPOSITION AND POTENCY ASSUMPTIONS

(UNITS ARE EQUIVALENT % SUCROSE)

Synergy

actual sweetness detected

○ Potencies at sweetness level equal to 10 % sucrose.
+ Potencies at actual Percent concentration.

RATIO OF ASPARTAME : SODIUM SACCHARIN (WT : WT)

FIG. 15.5. SYNERGISM OF ASPARTAME WITH SODIUM SACCHARIN IN A SIMPLE WATER SYSTEM

sweetener, even when other sweet substances are added to raise the total sweetness of the system. It is uncertain whether potency which is ultimately determined by neurophysiology responds to individual chemical concentrations, or to the sum of sweet stimuli. Taking this approach the previous equation would be:

$$S = C_A \times 133 + C_s \times 200$$

where: 133 is the potency of aspartame versus 10% sucrose and 200 is the potency of saccharin versus 10% sucrose, since 10% sucrose is the sweetness level of the system.

This approach naturally will show more synergism than the traditional method and is depicted as the (0) points in Fig. 15.5. The two lines should define the limits within which sweetness expectations should lie. At either extreme synergism is proven, in water, over a range of aspartame:saccharin ratios.

This mixture of sweeteners was also proposed by others (Hill and La Via 1972).

Applications

As shown in Table 15.2, the bulk compound has been evaluated in a wide variety of food products. In addition, it has been formulated as a tablet and as a spoon-for-spoon table top sweetener. A sampling of studies follows:

Aspartame was evaluated in frozen dessert by substituting it directly for saccharin in a dietetic ice milk formulation (0.03% aspartame substituted for 0.0085% saccharin). When the ice milk product was stored at -16°F for 6 months, it was found to be quite acceptable and was preferred for sweetness by 40% of the panel. After 6 months at -16°F, the development of crystallization became significant in both samples but less with the sucrose-containing samples. When the ice milk product was placed in storage at +10°F, the product retained overall acceptability for up to 6 weeks. At 6 weeks, one-third of the panel preferred the aspartame product for sweetness, despite the fact that it was significantly more "sandy" than the sucrose control. After 8 weeks, even the sucrose control began to show loss of smoothness.

Analytical data show that, after 6 months at -16°F, 88% of the original aspartame remained; at +10°F, after 6 weeks storage, 95% of the original aspartame remained (based upon the detection of the conversion product).

A spoon-for-spoon table top product was formulated using aspartame and was compared to a commercial spoon-for-spoon saccharin product and to sugar. During any one comparison a judge would be exposed to only 2 of the 3 samples on a random basis; and each sample was presented alone, not side-by-side. The tests were made in hot coffee and in cold cereal systems and tested in a presweetened (preprepared) and self-sweetened (self-prepared) condition. Paired preferences were developed from the response to the question: "All things considered, which sweetener did you prefer: sweetener served first; sweetener served second; or no preference?" These data are tabulated in Table 15.5. The paired preferences in all cases show a trend toward aspartame over the saccharin product and in pre-sweetened cereal they show a preference even over sucrose.

TABLE 15.5

PERCENT PREFERENCE FOR PAIRED COMPARISONS OF TABLE TOP SWEETNESS

	Coffee			Cereal		
	Aspartame-Based	Saccharin-Based	Sucrose	Aspartame-Based	Saccharin-based	Sucrose
Pre-	53.8	46.2		62.3	37.7	
sweetened	39.5		60.5	59.8		40.2
		35.9	64.1*		47.4	52.6
Self-	51.9	48.1		63.3*	36.7	
sweetened	21.2		78.8*	36.9		63.1*
		19.9	80.1*		25.3	74.7*

*Significance at the 90% level

Another benefit of the spoon-for-spoon form of the sweetener, which is apparent in cereal and on fruit, relates to the dissolving rate. Sugar in cereal tends to settle in the bottom of the bowl, undissolved, whereas the spoon-for-spoon form of aspartame dissolves. On fruit, another benefit is observed in the form of surface sweetness created on the fruit due to the better sticking and dispersion of the sweetener, or perhaps to a potentiation effect.

A fascinating problem developed in determining the correct aspartame content for the tablet form of the sweetener, since a tablet constitutes an integer dose. If one 20-mg tablet is not sweet enough, the next step is 2 tablets at 20 mg. The question was: how much aspartame should be used in formulating tablets which would be used predominantly in coffee? The requirements were (1) that users must be satisfied with either one or two tablets and (2) that at least half the population must be satisfied with one tablet.

Possible responses were analyzed by the matrix in Table 15.6, and possible dissatisfactions were classed into three categories: one tablet is too sweet; two tablets are not sweet enough; or one tablet is not sweet enough, *but* two tablets are too sweet.

TABLE 15.6

ANALYSIS OF ADVERSE CONSEQUENCES OF A GIVEN LEVEL OF ASPARTAME IN
TABLET FORMULATION

| | | One Tablet | |
	Too Sweet	Just Right	Not Sweet Enough
	(a)		(c)
Too sweet	Dissatisfaction: Must break tablet or quit	Acceptance	Dissatisfaction: must break tablet or quit
Just right	(Inconsistent)	Acceptance	Acceptance
Not sweet enough	(Inconsistent)	(Inconsistent)	(b) Dissatisfaction: must use ≥ 3 or quit

(Left side label: Two Tablets)

Since the tablets were yet to be prepared, the study was performed with coffee which had been presweetened with different levels of unformulated aspartame.

Based upon the data already in hand from Fig. 15.2, 5 sweetness levels of aspartame-presweetened coffee were selected for the test. Panelists were randomly given one of the middle 3 levels and asked to indicate if the sweetness level was over-sweetened, just right, or under-sweetened. The same panelists were then selected again for a second judgment, but this time the level offered was randomized within 3, based upon the first judgment. For example, if the first level was too sweet, the second level offered would be randomly selected from the lowest 3 of the 5 sweetness levels, since the panelist had already rejected high levels. The data thus developed, from about 200 panelists, were then statistically reduced to the form shown in Fig. 15.6. The 3 dissatisfactions [(a), (b), and (c)] are shown, as is the curve representing their sum. The minimum in the summation curve is the recommended dose level, based on satisfaction alone.

OTHER ASPARTIC ACID-BASED SWEETENER STUDIES

Some of the compounds discussed earlier have shown sufficient promise, based upon preliminary screening, to warrant a formal taste evaluation and those compounds and the taste results are shown in Table 15.7.

Other compounds are still being reviewed and introduced for further evaluation.

TABLE 15.7

TASTE CHARACTERISTICS OF OTHER PROMISING SWEETENERS

Sweetener	Water System		Food System by method of Fig. 15.1		Comments
	% Sucrose	Potency	% Sucrose	Potency	
L-Aspartyl-L-tyrosine methyl ester	0.34*	100	Coffee-4%	60-65	No consistent character difference from sucrose in coffee. Some bitterness detected at highest concentration.
	4.3	73			
	10.1	25			
	14.6	15			
N-(D,L-1, 4-Dimethyl-pentyl)-α-L-aspartamide	0.34*	300			
	4.3	115			
	10.1	63			
	14.6	49			
N-(L-1,4-Dimethylpentyl)-α-L-aspartamide			Coffee-4%	170-180	No consistent character difference from sucrose in coffee, but in orange pop an uncharacterized after-taste was noted.
			Orange 16% Pop	100-105	
L-Aspartyl-L-(3-cyclohexyl)alanine, methyl ester			Coffee-4%	210-240	
			Orange 16% Pop	130-160	

*Recognition Threshold

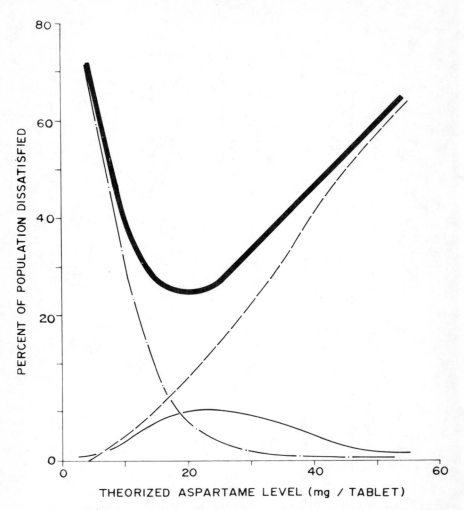

FIG. 15.6. EFFECT OF ASPARTAME CONTENT ON PREDICTED TABLET ACCEP-
TANCE (BASED UPON A COFFEE MODEL)— — one tablet is too sweet —·— two tablets
are not sweet enough, —— one is not enough but two are too much ▬▬ summation of the
three dissastisfaction curves

ACKNOWLEDGMENT

Data presented in this report represents the efforts of many
individuals. In particular, I would like to recognize John Colliopoulos
and Louis Jokay in the area of food technology, and Dave Calhoun
in the area of statistics. Dr. G. Anthony provided analytical data.
Some of the data presented here were performed by contract.

BIBLIOGRAPHY

AMERINE, M. A., PANGBORN, R. M., and ROESSLER, E. B. 1965. Principles of sensory evaluation of foods. Academic Press, New York.

CAUL, JEAN F. 1957. The Profile method of flavor analysis. Advances in Food Research 7, 1-40.

CLONINGER, M. R., and BALDWIN, R. E. 1970. A low-calorie sweetener. Science 170, 81-82.

DAVEY, J. M., LAIRD, A. H., and MORLEY, J. S. 1966. Polypeptides. Part III. The synthesis of the C-terminal tetrapeptide sequence of gastrin, its optical isomers, and acylated derivatives. J. Chem. Soc. p. 555.

FINNEY, D. J. 1947. Probit Anlysis, Cambridge Univ. Press.

HILL, J. A., and LA VIA, A. L. 1972. Sweetening compositions containing saccharin and dipeptides. U.S. 3,695,898, Oct. 3.

MAZUR, R. M., SCHLATTER, J. M., and JOKAY, L. 1969. New sugar substitute. Food Technol. 23, 4, 508.

MAZUR, R. H., and CRAIG, T. 1970. A new sugar substitute. Proc. Soc. Soft Drink Technol. 57.

MORTON, L. 1972. Comparative sweetness of different sugars. Proc. Soc. Soft Drink Technol., 77-86.

MOSKOWITZ, H. R. 1970A. Sweetness and intensity of artificial sweeteners. Percept. and Psychophys. 8, No. 1, 40-42.

MOSKOWITZ, H. R. 1970B. Ratio scales of sugar sweetness. Percept. and Psychophys. 7, No. 5, 315-320.

MOSKOWITZ, H. R. 1971. Intensity scales for pure taste and for taste mixtures. Percept. and Psychophys. 9, No. 1A, 51-56.

SALANT, ABNER 1968. Non-nutritive sweeteners. In Furia, T. E. Handbook of Food Additives. Chemical Rubber Co., Cleveland, O.

STEVENS, S. S., 1956. The direct estimation of sensory magnitudes—loudness. Am. J. Psychol. 69, 1.

STONE, HERBERT, and OLIVER, SHIRLEY M. 1969. Measurement of the relative sweetness of selected sweeteners and sweetener mixtures. J. Food Sci., 34, 215-222.

CHAPTER 16

R. M. Horowitz
Bruno Gentili

Dihydrochalcone Sweeteners

Relatively few naturally occurring dihydrochalcones are known and these have only a limited distribution in plants (Williams 1966). The dihydrochalcone phloridzin (1) is the characteristic bitter flavonoid of apple trees. Another dihydrochalcone is glycyphyllin (2), isolated from a sweet-tasting Australian shrub, *Smilax glycyphylla*, by Wright and Rennie (1881) and described as " . . . only very sparingly soluble in cold water, but sufficiently so to communicate its characteristic strong licorice-like taste" (Rennie 1886). In our experience, however, glycyphyllin seems bitter-sweet, and the element of bitterness predominates.

1 R = β-D-Glucosyl
2 R = α-L-Rhamnosyl

These facts about taste were unfamiliar to us in the late fifties when we first became interested in the structure-activity relations of citrus flavanones. We had found that the bitter citrus flavanone glycosides, 3-6, always contained the disaccharide neohesperidose (2-O-α-L-rhamnosyl-β-D-glucose) linked to the 7-hydroxy group of the aglycone. A second group of citrus flavanones, 7-10, was structurally the same in every detail except that they contained the isomeric disaccharide, rutinose (6-O-α-L-rhamnosyl-β-D-glucose). The transposition of rhamnose to the 6-position resulted in a complete loss of bitterness and gave rise to a series of tasteless compounds. Thus, we had to conclude that the point of attachment of rhamnose to glucose was of fundamental importance in determining bitterness or tastelessness in these compounds (Horowitz and Gentili 1963A).

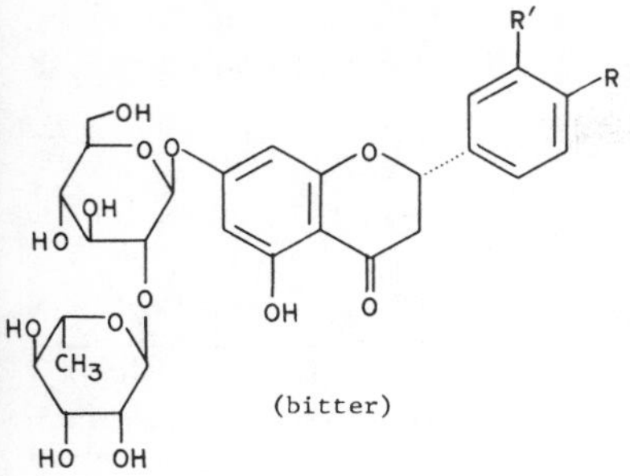

(bitter)

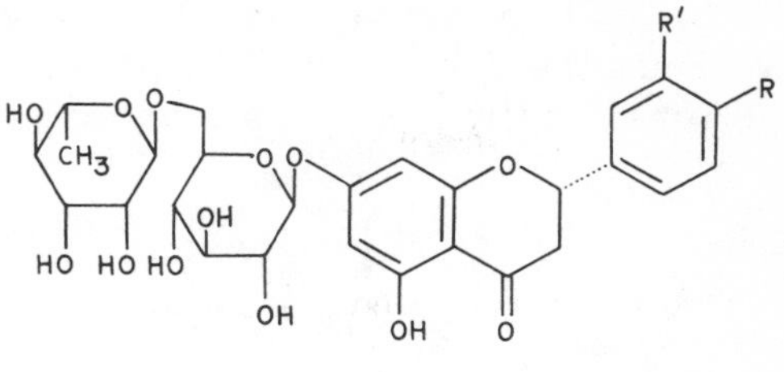

(tasteless)

3 R = OH, R' = H (naringin)
4 R = OCH$_3$, R' = OH (neohesperidin)
5 R = OCH$_3$, R' = H (poncirin)
6 R = R' = OH (neoeriocitrin)

7   R = OH, R' = H (naringenin rutinoside)
8   R = OCH$_3$, R' = OH (hesperidin)
9   R = OCH$_3$, R' = H (isosakuranetin rutinoside)
10 R = R' = OH (eriocitrin)

These observations prompted us to make various modifications of the bitter compounds to see whether we could map out the general structural requirements for bitterness. One modification was to convert the flavanone naringin to the corresponding chalcone by means of alkali, and then to the dihydrochalcone (11) by catalytic hydrogenation of the chalcone. The reactions take place readily:

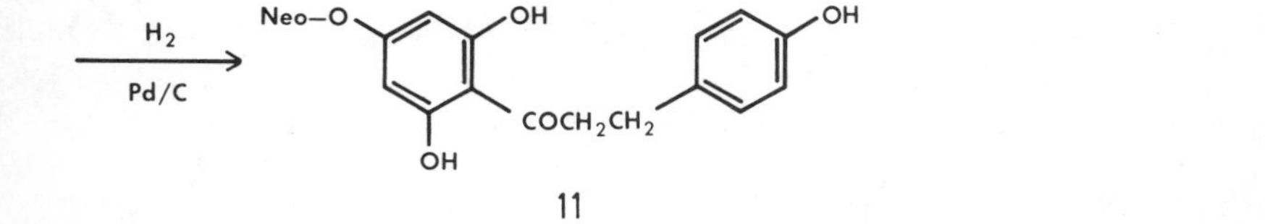

3

Naringin chalcone

11

We discovered that both the chalcone and dihydrochalcone are sweet, and moreover, that the sweetness more or less matched in intensity the extreme bitterness of the parent substance naringin. This unexpected result led us to study the conversion of the other bitter citrus flavanones to their respective dihydrochalcones. One of them, neohesperidin (4), yielded a dihydrochalcone (12) outstanding for its level of sweetness (Horowitz and Gentili 1963B) (Table 16.1). The others, poncirin (5) and neoeriocitrin (6), yielded dihydrochalcones that were bitter or only slightly sweet.

TABLE 16.1

TASTE AND RELATIVE SWEETNESS OF DIHYDROCHALCONES AND SACCHARIN

Compound	Taste	Relative sweetness (molar)	Relative sweetness (weight)
Naringin dihydrochalcone	Sweet	1	0.4
Neohesperidin dihydrochalcone	Sweet	20	7
Poncirin dihydrochalcone	Sl. bitter	—	—
Neoeriocitrin dihydrochalcone	Sl. sweet	—	—
Hesperetin dihydrochalcone glucoside	Sweet	1	0.4
Saccharin (Na)	Sweet	1	1

Included in Table 16.1 is hesperetin dihydrochalcone $4'$-O-β-D-glucoside (HDG) (13), which, like the dihydrochalcones 11 and 12, promises to be useful as a sweetener. It can be prepared most economically from the abundant flavanone hesperidin (8), which thus far has had few important uses:

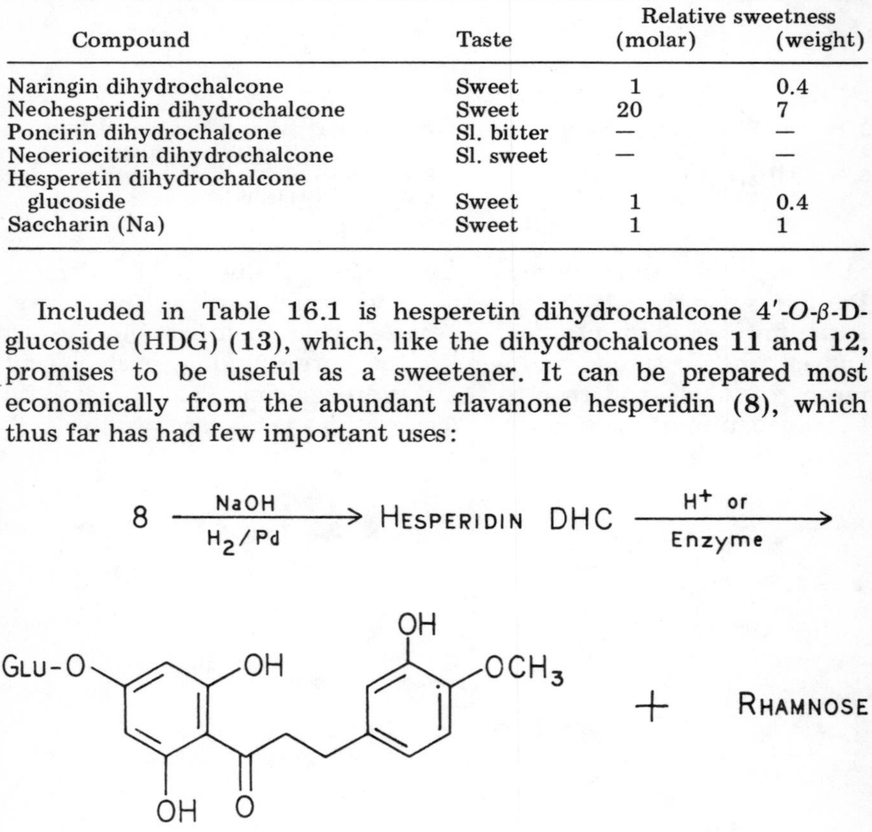

$$8 \xrightarrow[\text{H}_2/\text{Pd}]{\text{NaOH}} \text{HESPERIDIN} \ \ \text{DHC} \xrightarrow[\text{Enzyme}]{\text{H}^+ \text{ or}}$$

+ RHAMNOSE

13

The only step of possible difficulty in these reactions is the partial hydrolysis of hesperidin dihydrochalcone to remove rhamnose from the disaccharide while leaving glucose attached to the aglycone. This hydrolysis can be accomplished either by very dilute acid (Horowitz and Gentili 1969) or, more efficiently, by the enzyme rhamnosidase (Horowitz and Gentili 1971A).

Two points of interest are illustrated by these reactions.

1. Starting with the tasteless flavanone hesperidin one obtains the tasteless hesperidin dihydrochalcone. We have found here and in other cases that compounds which contain the intact disaccharide rutinose are invariably tasteless. Only when rhamnose is removed (as in **13**) do we obtain a sweet (or bitter) compound. Thus, rhamnose situated at the C-6 hydroxyl group of glucose abolishes taste responses.

2. Compound **12** differs structurally from compound **13** only in having rhamnose attached to the C-2 hydroxyl group, yet is about 20 times sweeter. In this and other examples we find that the occurrence of the 2-*O*-rhamnosyl group usually enhances the taste (sweet or bitter) that subsists in the simple glucoside.

The large-scale preparation of naringin dihydrochalcone from commercially available naringin is accomplished easily and in high yield by catalytic hydrogenation in an alkaline solution. The same would apply to neohesperidin dihydrochalcone (**12**) except that the starting material, neohesperidin, which occurs in the sparsely cultivated Seville or bitter orange (*Citrus aurantium*), is not a commercial product. It is necessary, therefore, to obtain **12** by the conversion of naringin (**3**). This is done (Fig. 16.1) by cleaving

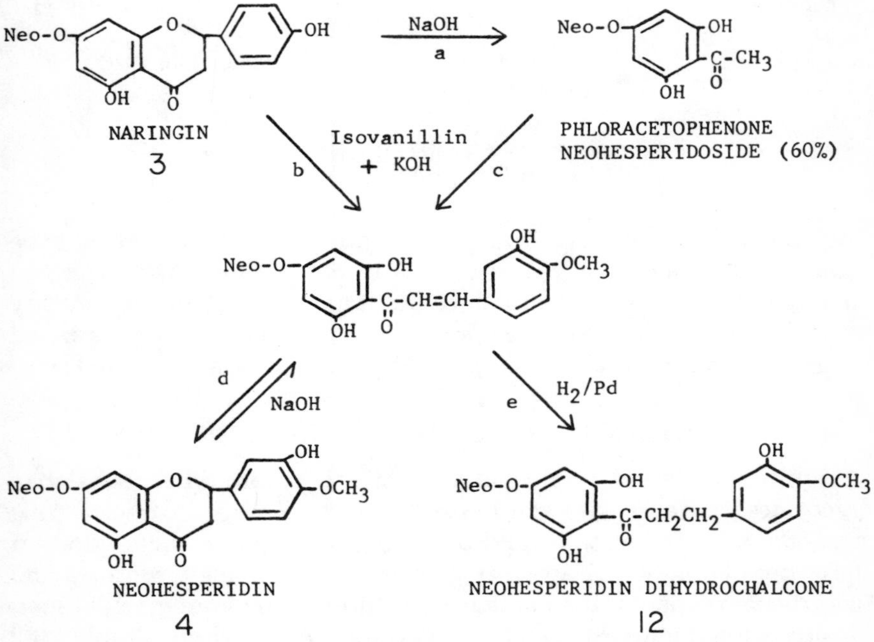

FIG. 16.1. CONVERSION OF NARINGIN TO NEOHESPERIDIN DIHYDROCHALCONE

naringin with alkali to give phloracetophenone 4'-O-β-neohesperidoside (path a) which is then condensed with isovanillin to give neohesperidin chalcone and neohesperidin (c,d). Alternatively, the naringin can be treated directly with isovanillin and alkali to give the chalcone and flavanone (b,d) without isolating any intermediates (Horowitz and Gentili 1968). In either case the final product is obtained by hydrogenation. Although sizable quantities of 12 have been produced, the condensation of phloracetophenone neohesperidoside with isovanillin requires careful attention to detail in order to obtain optimum yields, which are ~ 30% based on naringin (Robertson *et al.* 1973).

Figure 16.2 summarizes the preparation and interconversions of dihydrochalcones 11, 12 and 13.

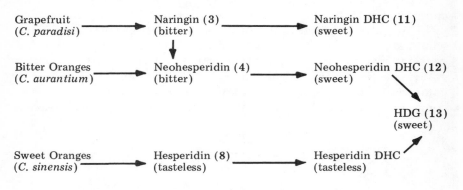

FIG. 16.2. ORIGIN AND INTERCONVERSION OF
DIHYDROCHALCONES

The preparation of 13 from 12, although easy, would not be economical. Since the sweet orange is by far the most widely grown citrus crop in the world, and since hesperidin is available in quantity as a by-product of the orange-processing industry, sweeteners based on hesperidin are attractive subjects for further research.

STRUCTURE-ACTIVITY RELATIONS

During the last 10 yr a large number of new dihydrochalcone glycosides have been synthesized in the hope of clarifying structure-activity relations and obtaining useful sweeteners. Most of these new compounds are B-ring variants of existing structures and have been prepared by condensing phloracetophenone neohesperidoside with the appropriately substituted benzaldehyde. Table 16.2 lists these derivatives and their taste. The data in the table have been

TABLE 16.2

STRUCTURE AND TASTE OF DIHYDROCHALCONE 4'-*O*-β-NEOHESPERIDOSIDES MODIFIED IN THE B-RING

| Compound | Substituent present at | | | | Taste[a] |
	2	3	4	5	
Mono- or polyhydroxy-substituted					
11			OH		+++
14[b]		OH			+++
15[b]	OH				—
16		OH	OH		+
17		OH	OH	OH	Nil
Hydroxy-alkoxy-disubstituted					
12		OH	OMe		+++
18[b]		OH	OEt		+++
19[b]		OH	O-n-Pr		++++
20		OH	O-i-Pr		++
21	OH	OMe			++
22		OMe	OH		Nil
23		OEt	OH		Nil
Hydroxy-alkoxy trisubstituted					
24		OH	OMe	OH	Nil
25	OMe	OH	OMe		Nil
26		OH	OMe	OMe	Nil
No hydroxyl substituent					
27			OMe		-+
28		OMe	OMe		-+
29		Me	OMe		Nil

[a]+ signifies sweet; — signifies bitter.
[b]Data of Krbechek *et al.* 1968.

discussed elsewhere in detail (Horowitz and Gentili 1971B). The significant conclusions with regard to B-ring substitution are as follows:

1. A B-ring hydroxyl group is needed for sweetness (11, 14, 16, 12, 18, 19, 20, 21), but its mere presence does not guarantee sweetness (15, 17, 22, 23, 24, 25, 26).

2. The absence of a hydroxyl group assures nonsweetness (29) or bitter-sweetness (27, 28).

3. For sweetness in hydroxy-alkoxy disubstituted compounds, the order of the substituent groups must be R-H-OH-Alkoxy (12, 18, 19, 20) or R-OH-Alkoxy (21), where R is the substituent at C-1 of the B-ring.

4. Taste is abolished if the order of groups in hydroxy-alkoxy disubstituted compounds is R-H-Alkoxy-OH (22, 23).

5. Taste is abolished if three adjacent groups are present in addition to R (24, 25, 26).

These conclusions should be useful as guides in choosing new derivatives to be synthesized.

A second group of variants is represented by compounds in which the hydroxyl groups of the A-ring have been alkylated. These are shown in Table 16.3.

TABLE 16.3

STRUCTURE AND TASTE OF DIHYDROCHALCONE 4'-O-β-NEOHESPERIDOSIDES
MODIFIED IN ONE OR BOTH
RINGS

| Compound | Substituent present at | | | | Taste[a] |
	$2'$	$6'$	3	4	
30	OMe	OH	—	OH	+
31	OMe	OH	OH	OMe	++
32	OEt	OH	OH	OMe	+
33	OMe	OH	OMe	OMe	—+
34	$OCH_2 COOH$	OH	OH	OMe	Nil
35	OH	OH	OH	$OCH_2 COOH$	Nil
36[b, c]	OH	—	OH	OMe	+[d]
37[b, c]	OH	—	OH	OEt	+[d]

[a]+ signifies sweet; — signifies bitter.

[b]Farkas and Nogradi 1972.

[c]β-D-glucosyloxy at position 4'.

[d]No quantitative data given.

In general, alkylation of an A-ring hydroxyl has a deleterious effect on both sweetness and solubility (**30, 31, 32**). We have encountered no examples where A-ring alklyation increases or even maintains the existing sweetness. Compound **33** has an element of bitterness, which is to be expected since it contains no B-ring hydroxyl (compare with **28**). Compounds **34** and **35** were synthesized in order to obtain dihydrochalcones with improved solubility. Neither, however, was appreciably sweet. Compounds **36** and **37**, in which the 6'-hydroxyl substituent is absent, have been claimed as sweetening agents in a recent patent (Farkas and Nogradi 1972).

The specific structural features required for sweetness in the sugar component of dihydrochalcone glycosides have not been studied in much detail. Unfortunately, syntheses designed to produce glycosidic bonds are often tedious and afford low yields. In general, the taste responses produced by dihydrochalcone glucosides and neohesperidosides (2-O-rhamnosylglucosides) are qualitatively similar, although the response is usually enhanced in the neohesperidoside. Taste is abolished in rutinosides (6-O-rhamnosylglucosides) but is retained in 6-O-methylneohesperidosides. For example, the 6''-O-methyl ether of neohesperidin dihydrochalcone (**38**) is indistinguishable in taste from the parent compound, neohesperidin dihydrochalcone (**12**).

38

Thus, a bulky sugar group (rhamnose) enhances sweetness when linked to the C-2 hydroxyl of glucose, but abolishes sweetness when linked to the C-6 hydroxyl. A small group (methyl) apparently has little effect on taste when situated at the C-6 hydroxy. It is evident that neither a free C-2 nor C-6 hydroxyl group is essential, as such, to impart sweetness, and it may even be posssible to eliminate these substituents entirely and still retain the taste response. We have recently synthesized hesperetin dihydrochalcone 4'-O-β-D-xyloside (39), the xylose analog of HDG (13). This compound, in which the C-6 hydroxyl as well as the adjacent —CH_2— group are absent, was found to be about twice as sweet as HDG.

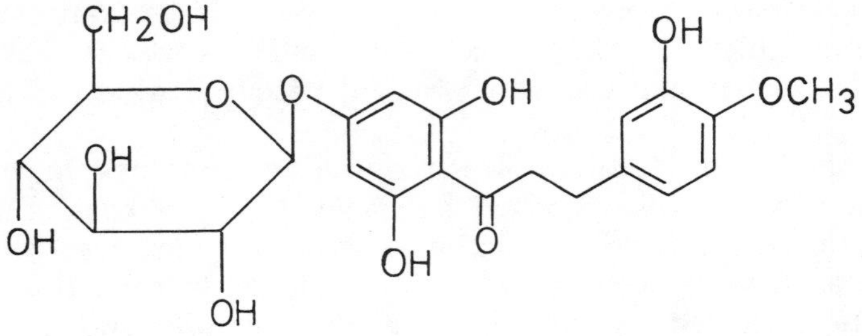

13

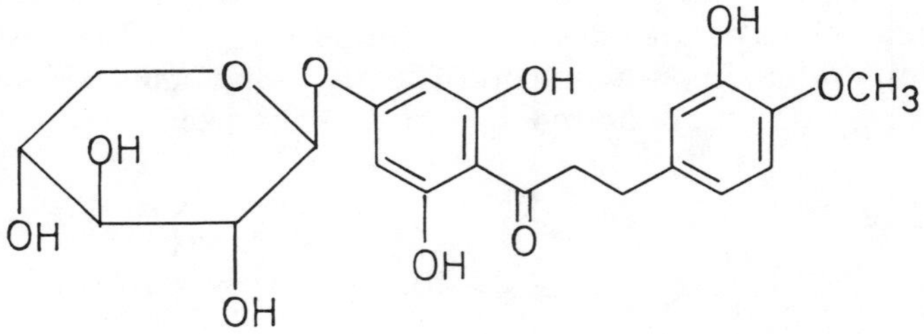

39

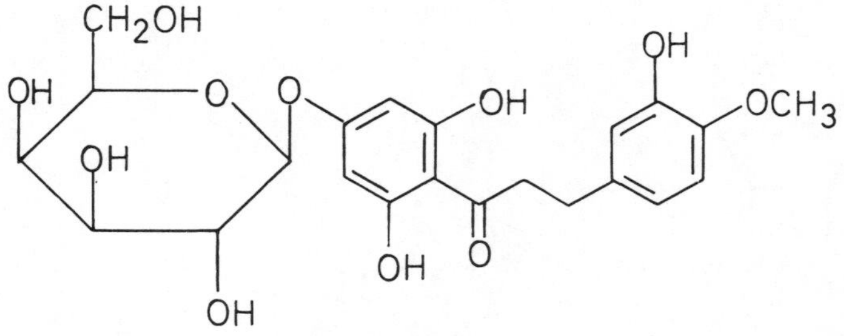

40

The evidence suggests that the C-3 and C-4 hydroxyl groups are intimately involved in the taste response, and this idea accords with results obtained by Evans (1963) on the response of insects to sugars.

We might anticipate that methylation or epimerization of either of these hydroxyl groups would have a marked effect on taste. Recently we prepared hesperetin dihydrochalcone 4'-O-β-D-galactoside (40) and found it to be ~ 1.5 to 2 times sweeter than HDG. These compounds differ only in the stereochemistry of the C-4 hydroxyl group of the sugar. It is of interest that van Niekerk and Koeppen (1972) synthesized the 2-O-rhamnosylgalactoside of naringenin dihydrochalcone (41) and found it about equal in sweetness to the 2-O-rhamnosylglucose analog, naringin dihydrochalcone (11).

The structure of a new natural product, osladin (42), was determined recently by Sorm's group (Jizba *et al.* 1971). Osladin is a

saponin-like sweet principle in the rhizomes of the fern *Polypodium vulgare*. Although the disaccharide component is shown in the original paper as 2-O-β-L-rhamnosyl-β-D-glucose and it is stated that the configuration of the rhamnose is unknown, it seems likely that it must be 2-*O*-α-L-rhamnosyl-β-D-glucose, *i.e.*, neohesperidose. This provides still another example of an intensely sweet compound that contains a 1′ → 2 linked disaccharide bound glycosidically to an aglycone. Others are stevioside, glycyrrhizin and, of course, the dihydrochalcones (see Chapter 20).

APPLICATIONS

Dihydrochalcones such as **11, 12, 13, 39** and **40** are characterized by a pleasant sweetness, somewhat slow in its onset and of varying (usually long) duration. There is little, if any, bitter after-taste, but a sensation somewhat reminiscent of that given by licorice or menthol occurs. It seems likely that the initial applications of these compounds will be in products such as chewing gums, pharmaceuticals, mouthwashes, toothpastes, etc., where long-lasting sweetness and ability to mask bitterness are an advantage. Other foreseeable uses are in beverages and chocolate.

Combinations of saccharin and neohesperidin dihydrochalcone are said to be synergistic, according to a recent patent (Ishii *et al.* 1972). The use of the sodium and calcium salts of **11** and **12**, respectively, has been advocated because of increased solubility (Westall and Messing 1972). In another patent (Rizzi and Neely 1972), solutions of the aglycone hesperetin dihydrochalcone in solvents such as ethanol or propylene glycol have been recommended as sweetening and flavoring agents. It had been shown earlier that the aglycone is moderately sweet, but its solubility in water is slight (Horowitz 1964).

Thin-layer and paper chromatographic systems for separating dihydrochalcone glycosides from each other and from their precursors have been described (Gentili and Horowitz 1971). The same paper describes a reagent—sodium borohydride/dichlorodicyanobenzoquinone/*p*-toluenesulfonic acid—which reacts with many dihydrochalcones to give red-purple colorations. Since the reagent appears to be specific for dihydrochalcones, it should be useful in the analysis of complex mixtures containing these substances.

Finally, it must be mentioned that the all-important two-year feeding studies required before dihydrochalcones can be considered for general use are now being carried out at the Western Regional Research Laboratory by Dr. A. N. Booth, who is using neohesperidin dihydrochalcone in these experiments.

BIBLIOGRAPHY

EVANS, D. R. 1963. Chemical structure and stimulation by carbohydrates. *In* Olfaction and Taste, Y. Zotterman (Editor). The Macmillan Co., New York.

FARKAS, L. and NOGRADI, M. 1972. Dihydrochalcone sweetening agents. Hung. Teljes 4026. Chem. Abstr. 77, 60321 (1972).

GENTILI, B. and HOROWITZ, R. M. 1971. Chromatography of dihydrochalcone sweeteners and related compounds. A reagent for detecting dihydrochalcones. J. Chromatog. 63, 467-469.

HOROWITZ, R. M. 1964. Relations between the taste and structure of some phenolic glycosides. In Biochemistry of Phenolic Compounds, J. B. Harborne (Editor). Academic Press, New York.

HOROWITZ, R. M. and GENTILI, B. 1963A. Flavonoids of citrus. VI. The structure of neohesperidose. Tetrahedron 19, 773-782.

HOROWITZ, R. M. and GENTILI, B. 1963B. Dihydrochalcone derivatives and their use as sweetening agents. U.S. Patent 2,087,821. April 30.

HOROWITZ, R. M. and GENTILI, B. 1968. Conversion of naringin to neohesperidin and neohesperidin dihydrochalcone. U.S. Patent 3,375,242. March 26.

HOROWITZ, R. M. and GENTILI, B. 1969. Preparation of hesperetin dihydrochalcone glucoside. U.S. Patent 3,429,873. February 25.

HOROWITZ, R. M. and GENTILI, B. 1971A. Enzyme preparation of hesperetin dihydrochalcone glucoside. U.S. Patent 3,583,894. June 8.

HOROWITZ, R. M. and GENTILI, B. 1971B. Dihydrochalcone sweeteners. In Sweetness and Sweeteners, G. G. Birch, L. F. Green and C. B. Coulson (Editors). Applied Science Publishers, London.

ISHII, K., TODA, J., AOKI, H. and WAKABAYASHI, H. 1972. Sweetening composition. U.S. Patent 3,653,923. April 4.

JIZBA, J., DOLEJS, J., HEROUT, V. and SORM, F. 1971. The structure of osladin—the sweet principle of the rhizomes of Polypodium vulgare. Tetrahedron Letters 1329-1332.

KRBECHEK, L. et al. 1968. Dihydrochalcones. Synthesis of potential sweetening agents. J. Agr. Food Chem. 16, 108-112.

VAN NIEKERK, D. M. and KOEPPEN, B. H. 1972. Synthesis of 2-O-α-L-rhamnopyranosyl-D-Galactose, a reported partial hydrolysis product of α-solanine, and some taste-eliciting flavonoid 2-O-α-L-rhamnopyranosyl-β-D-galactopyranosides. Experientia 28, 123-124.

RENNIE, E. H. 1886. Glycyphyllin, the sweet principle of Smilax glycyphylla. J. Chem. Soc. 49, 857-865.

RIZZI, G. P. and NEELY, J. S. 1972. Hesperetin dihydrochalcone sweeteners. Ger. Offen. 2,148,332. Chem. Abstr. 77, 86777 (1972).

ROBERTSON, G. H., CLARK, J. P. and LUNDIN, R. 1973. Dihydrochalcone sweeteners: production of neohesperidin dihydrochalcone. I & EC Prod. Res. & Development (in press).

WESTALL, E. B. and MESSING, A. W. 1972. Monobasic calcium and sodium salts of neohesperidin dihydrochalcone. Ger. Offen. 2,216,071. Chem. Abstr. 78, 43051 (1973).

WILLIAMS, A. H. 1966. Dihydrochalcones. In Comparative Phytochemistry, T. Swain (Editor). Academic Press, London.

WRIGHT, C. R. A. and RENNIE, E. 1881. Note on the sweet principle of Smilax glycyphylla. J. Chem. Soc. 39, 237-240.

CHAPTER 17

H. van der Wel | # Miracle Fruit, Katemfe, and Serendipity Berry

Since the ban on cyclamates, there has been an increasing interest in alternative low-caloric sweetening agents. Nature is the most logical starting point for any search for nontoxic sweeteners. Although natural occurrence does not warrant nontoxicity, we looked for plant material consumed by various indigenous people, as on reasonable grounds these can be assumed to be nontoxic.

A fruit with unusual taste properties, which has attracted the attention of many investigators, is miracle fruit or miraculous berry, the fruit of *Synsepalum dulcificum* (Schum. et Thonn.). It has the unique property of causing sour materials to taste sweet after the mouth has been exposed to the fruit pulp. The berries of the plant, found in tropical West Africa, are red, 1 to 2 cm long, olive-shaped and consist of a relatively large seed surrounded by a thin layer of fruit pulp.

Inglett and May (1968), who screened the tropical vegetation of Africa for fruits with a sweet taste, selected two plants for further research, *Thaumatococcus daniellii* Benth and *Dioscoreophyllum cumminsii* Diels. The fruit of Thaumatococcus is found in West Africa, where it is called Katemfe by the natives. A single fruit weighing about 16 gm comprises three triangular fleshy pericarps, each of which contains a large black seed surrounded by a thick transparent mucilage with a soft, jelly-like aril covering the base of the seed. These arils have an intense sweet taste and a slight licorice after-taste with some cooling effect. They are used by the natives to sweeten palm wine and fruit drinks. The sweet principle is water-soluble (Inglett and May 1968).

The fruit of *Dioscoreophyllum cumminsii* or Serendipity berry is found in several regions of tropical Africa. The round red berries, approximately 1 cm in diameter, grow in grape-like clusters with 50 to 100 berries in each bunch. The fruit pulp contains a water-soluble, intensely sweet-tasting compound (Inglett and May 1969). In the Congo the fruit is eaten by the natives.

MIRACLE FRUIT

Inglett *et al.* (1965) made the first trials to isolate the sweetness-inducing principle of the fruit of *Synsepalum*, but they were unsuccessful in solubilizing it, either by extraction with water, salt solutions and organic solvents or by treatment with enzymes. Brouwer *et al.* (1968) succeeded in extracting the active principle by using whole human saliva of pH $\geqslant$ 7. Later it was found that similar results could be obtained with aqueous solutions of neuraminic acid derivatives, tannin-binding agents like polyethylene glycol, gelatin or casein at pH $\geqslant$ 7. Gel filtration indicated a molecular weight of the order of 10^6, and starch-gel and paper electrophoresis suggested a neutral or weakly acidic entity. Much better results could be obtained by extracting the berries with an aqueous solution of strongly basic compounds like salmine, a protamine, or naturally occurring polyamines like spermine. Purification of the extract by ammonium sulfate fractionation and successive gel filtration on Sephadex G-50 and G-25 yielded an almost pure preparation, which we called miraculin; it had a molecular weight of 42,000, as calculated from ultracentrifugation data, and an isoelectric point of about 9.

Incubation of miraculin with proteolytic enzymes destroyed the activity, indicating that the substance is a protein. This was confirmed by acid hydrolysis, which produced amino acids and sugars. Carbohydrate analysis revealed that miraculin contains 7.5% carbohydrate and is therefore a glycoprotein. Kurihara and Beidler (1968) confirmed these results. After solubilization in a carbonate buffer at pH 10.5, they found a molecular weight of 44,000 and 6.7% carbohydrate.

One hundred micrograms of miraculin is sufficient to change the sensation of sourness into sweetness for 1 to 2 hr. However, this long-lasting effect limits the use of miraculin as a sweetener. Moreover it is thermolabile and loses its activity at a pH below 2. Another limitation to practical application in food and drinks is its cumulative action. For instance, when miraculin is added to yoghurt, the yoghurt is sour at first, dependent on the concentration; but the more yoghurt one eats, the sweeter it becomes.

KATEMFE AND SERENDIPITY BERRY

The fruits of *Thaumatococcus* and *Dioscoreophyllum* were obtained from Nigeria. After the fruits were cut open, the arils of the *Thaumatococcus* fruit and the pulp of the Serendipity berries were collected. Both arils and pulp were extracted with water. According to the procedure indicated in Table 17.1, Katemfe yields two sweet-tasting proteins (van der Wel 1972; van der Wel and Loeve 1972), which we called thaumatin I and II. Using similar techniques we isolated a single sweet-tasting protein from the Serendipity berry (van der Wel 1972; van der Wel and Loeve 1973). Morris and Cagan (1972) from the Monell Chemical Senses Center achieved the same results independently and they called the sweet principle Monellin.

TABLE 17.1

ISOLATION PROCEDURE FOR THE SWEET-TASTING PROTEINS FROM
THAUMATOCOCCUS DANIELLII AND *DIOSCOREOPHYLLUM CUMMINSII*

(1) homogenization in water
(2) centrifugation
(3) ultrafiltration
(4) gel filtration
(5) desalting
(6) ion-exchange chromatography
(7) desalting
(8) freeze-drying

Isolation

The water-soluble components of the fruits were purified from low molecular weight material and concentrated by ultrafiltration. The concentrates were submitted to gel filtration. The elution profile of the *Thaumatococcus* concentrate (Fig. 17.1) shows 3 fractions having a sweet taste (A, B and C). Gel chromatography of the crude

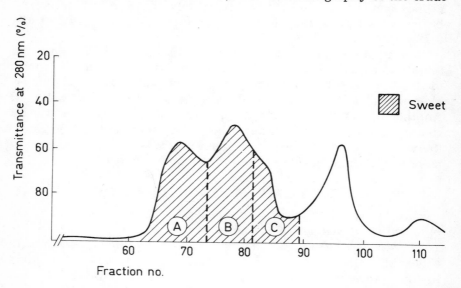

FIG. 17.1. GEL FILTRATION ON SEPHADEX G-50 OF 800 MG OF THE FREEZE-DRIED EXTRACT FROM *THAUMATOCOCCUS DANIELLII*
Bed: 5 x 88 cm; flow rate: 72 ml per hr; fraction volume: 18 ml; eluant: distilled water.

concentrate of the Serendipity berries on Sephadex G-100 reveals 3 peaks, only one of which had a sweet taste (Fig. 17.2a).

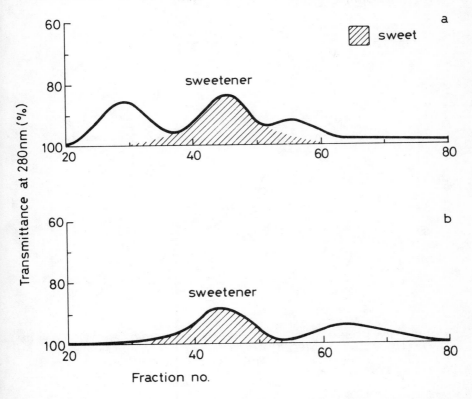

FIG. 17.2. GEL FILTRATION ON SEPHADEX (EQUILIBRATED WITH 0.1*M* NaCl) OF THE EXTRACT FROM *DIOSCOREOPHYLLUM CUMMINSII*

(a) 25 mg crude berry extract in 3.5 ml 0.1*M* NaCl on Sephadex G-100; bed: 1.5 x 38 cm; flow rate: 13.5 ml/hr; fraction volume: 1.35 ml.

(b) 7 mg of the G-100 fraction in 1.5 ml 0.1*M* NaCl on Sephadex G-50; bed: 1 x 90 cm; flow rate: 3.75 ml per hr; fraction volume 0.75 ml.

Rechromatography of this fraction on Sephadex G-50 gave rise to 2 peaks, the first of which presented a very sweet taste (Fig. 17.2b). For a further purification, the sweet-tasting fractions of both fruits were submitted to ion-exchange chromatography on SE-Sephadex C-25 with a linear sodium chloride gradient. In this way, 2 pure preparations, thaumatin I and thaumatin II, were obtained from the sweet-tasting gel filtrates from *Thaumatococcus* (fractions A, B and C, see Fig. 17.3) and one, Monellin freed from all impurities, from *Dioscorephyllum* (Fig. 17.4a). The purity of the fractions was checked by polyacrylamide-gel electrophoresis, using Amido Black 10 B as staining agent (Fig. 17.4b and 17.5); their protein content was assessed by the biuret method.

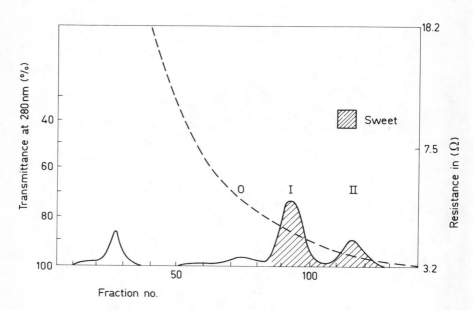

FIG. 17.3. ION-EXCHANGE CHROMATOGRAPHY OF 50 MG OF FRACTION A FROM *THAUMATOCOCCUS DANIELLII* IN 0.02*M* PHOSPHATE BUFFER pH 7.0 ON SE-SEPHADEX C-25 USING A LINEAR SALT GRADIENT
Bed: 1.5 x 91.8 cm; flow rate: 18 ml per hr; fraction volume: 2.2 ml.＿＿＿＿＿，transmittance;. . . . ., resistance.

Characterization

The protein character of the isolated sweet principles was proved by:

1. the presence of almost 100% polypeptide material as determined by the biuret method;

2. the yield of 100% amino acid on acid hydrolysis;

3. the characteristic ultraviolet absorption spectra;

4. their binding with Amido Black 10 B;

5. the disappearance of the sweet taste after incubation with trypsin.

The amino acid composition was determined by hydrolysis with 6*N* HCl at 110°C, at 4 hydrolysis times, using norleucine as internal standard and corrected by extrapolation to hydrolysis time t=0 (see

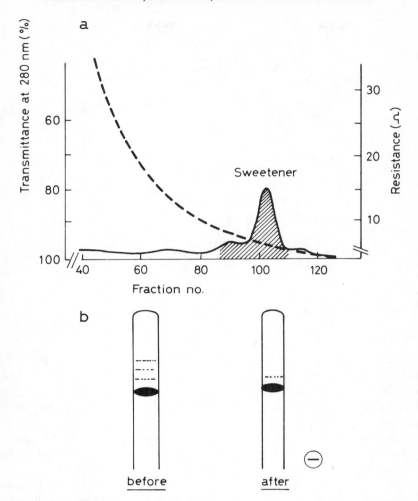

FIG. 17.4. (A) ION-EXCHANGE CHROMATOGRAPHY OF 26 MG OF THE G-50 FRACTION FROM *DIOSCOREOPHYLLUM CUMMINSII* IN 0.02*M* TRIS/HCl BUFFER pH 7.65 ON SE-SEPHADEX C-25, USING A LINEAR SALT GRADIENT

Bed: 1.5 x 61 cm; flow rate: 8.2 ml per hr; fraction volume: 1.6 ml._____, transmittance; .... resistance.

(b) Polyacrylamide-gel electrophoresis of the purified protein before and after ion-exchange chromatography.

Table 17.2). Histidine is not present in any of these proteins. Proton magnetic resonance measurements demonstrated the absence of fully methylated amino groups (Korver *et al.* 1973). Morris *et al.* (1973) were also unable to detect mono- or dimethyl derivatives of lysine or

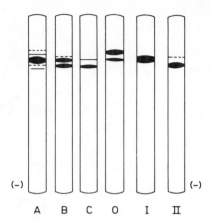

FIG. 17.5. POLYACRYLAMIDE-GEL ELECTROPHORESIS OF THE FRACTIONS A, B AND C OF THE GEL FILTRATION RUN SHOWN IN FIG. 17.1 AND OF PEAK O, I AND II OF THE GRADIENT CHROMATOGRAPHIC RUN SHOWN IN FIG. 17.3

arginine in Monellin. No free sulfhydryl groups were found in the proteins by the procedure, according to Ellman (1959). The excess of the free basic groups with respect to the number of free carboxylic groups is indicative of the basic character of these proteins.

The isoelectric point of Monellin was determined by preparative isoelectric focusing; those of the thaumatins were estimated by starch-gel electrophoresis at different pH values (see Table 17.3).

The molecular weights of the thaumatins were determined in an analytical ultracentrifuge, using the low-speed sedimentation equilibrium technique; the molecular weight of Monellin was estimated according to the method of Andrews (1965) by gel filtration on Sephadex G-50 (see Table 17.3). The extinction $A_{1\,cm}^{1\,\%}$ values were determined at 278 nm and at pH 5.6 (see Table 17.3). Sodium dodecyl sulfate polyacrylamide-gel electrophoresis, after cleavage of the disulfide bridges (Weiner et al. 1972) gave for all the proteins the same molecular weights as the techniques mentioned above, indicating that the proteins are single polypeptide chains. This has been confirmed by Morris et al. (1973) for Monellin. As there are no circular dichroism bands at 220 nm and 208 nm (characteristic of the α-helix), very little, if any, α-helix configuration is present in the molecule.

TABLE 17.2

AMINO ACID COMPOSITION (RESIDUES/MOL) OF THE SWEET-TASTING PROTEINS

Amino acid	Thaumatin I	Thaumatin II	Monellin
Aspartic acid	21	19	11
Glutamic acid	10	10	13
Serine	12	10	2
Threonine	19	17	4
Proline	12	12	6
Glycine	23	22	9
Alanine	15	14	3
Valine	9	8	4
Leucine	9	9	6
Isoleucine	7	7	7
Methionine	1	1	1
Phenylalanine	10	10	6
Tyrosine	7	8	5
Half-cystine	14	13	2
Histidine	—	—	—
Lysine	10	11	10
Arginine	11	12	8
Tryptophan	3	3	1
Ammonia	21	18	12
Total	193	186	98

TABLE 17.3

SOME CHARACTERISTICS OF THE SWEET-TASTING PROTEINS

Criteria	Thaumatin I	Thaumatin II	Monellin
Isoelectric point	12	12	9.03
Mol wt	21,000 ± 600	20,400 ± 600	11,500
$A_{1cm}^{1\%}$ (pH 5.6, 278 nm)	7.69	7.53	16.2
Sweetness intensity (times sweeter than sucrose)			
on a molar basis	1×10^5	1×10^5	8.4×10^4
on a weight basis	1600	1600	2500
Temperature (°C) above which sweetness disappears			
at pH 3.2	55	55	50
5.0	75	75	65
7.2	65	65	55

TASTE PROPERTIES OF THE PROTEINS

The sweetness intensity of the proteins as compared with sucrose was estimated in ranking tests (see Table 17.3).

Circular dichroism measurements show that the proteins undergo similar reversible conformational changes as the temperature is increased (see Fig. 17.6). At a certain temperature, however, varying

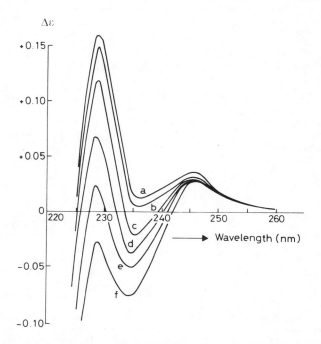

FIG. 17.6. CIRCULAR DICHROISM OF THAUMATIN I
(CONCN: 0.7 GM/LITER) AT PH 7.2
(a) 25-50°C;   (b) 62.5°C;   (c) 64°C;   (d) 65°C;   (e) 66°C;
(f) 67°C.

with pH, irreversible heat denaturation occurs (Table 17.3). Circular dichroism, proton magnetic resonance, and ultraviolet difference spectroscopy measurements show that tyrosine residues and at least 1 disulfidechromophore are involved in this conformational change. These irreversible changes coincide with a loss of sweetness, indicating that the groups underlying the conformational change are also operative in generating the sweet taste. Sweetness is also destroyed when the disulfide bridges are split, indicating the importance of the tertiary structure for the sweet taste. These findings and the complete loss of sweet taste on heating aqueous solutions above 75°C for a few minutes or on lowering the pH to below 2 at room temperature limit the use of these proteins as sweeteners.

BIBLIOGRAPHY

ANDREWS, P. 1965. The gel filtration behavior of proteins related to their molecular weight over a wide range. Biochem. J. *96*, 595-606.

BROUWER, J. N., VAN DER WEL, H., FRANCKE, A., and HENNING, G. J. 1968. Miraculin, the sweetness-inducing protein from Miracle fruit. Nature (London) *220*, 373-374.

ELLMAN, G. L. 1959. Tissue sulfhydryl groups. Arch. Biochem. Biophys. *82*, 70-77.

INGLETT, G. E., DOWLING, B., ALBRECHT, J. J., and HOGLAN, F. A. 1965. Taste-modifying properties of Miracle Fruit (*Synsepalum dulcificum*). J. Agr. Food Chem. *13*, 284-287.

INGLETT, G. E., and MAY, J. F. 1968. Tropical plants with unusual taste properties. Econ. Bot. *22*, 326-331.

INGLETT, G. E., and MAY, J. F. 1969. Serendipity berries—Source of a new intense sweetener. J. Food Sci. *34*, 408-411.

KORVER, O., VAN GORKOM, M., and VAN DER WEL, H. Spectrometric investigation of Thaumatin I and II, two sweet-tasting proteins from *Thaumatococcus daniellii* Benth. To be published.

KURIHARA, K., and BEIDLER, L. M. 1968. Taste-modifying protein from Miracle fruit. Science *161*, 1241-1243.

MORRIS, J. A., and CAGAN, R. H. 1972. Purification of Monellin, the sweet principle of *Dioscoreophyllum cumminsii*. Biochim. Biophys. Acta *261*, 114-122.

MORRIS, J. A., MARTENSON, R., DEIBLER, G., and CAGAN, R. H. 1973. Characterization of Monellin, a protein that tastes sweet. J. Biol. Chem. *248*, 534-539.

VAN DER WEL, H. 1972. Thaumatin, the sweet-tasting protein from *Thaumatococcus daniellii* Benth. *In* Olfaction and Taste. IV. D. Schneider (Editor). Wissenschaftliche Verlaggesellschaft MBH Stuttgart.

VAN DER WEL, H. 1972. Isolation and characterization of the sweet principle from *Dioscoreophyllum cumminsii* (Stapf) Diels. FEBS Lett. *21*, 88-90.

VAN DER WEL, H., and LOEVE, K. 1972. Isolation and characterization of Thaumatin I and II, the sweet-tasting proteins from *Thaumatococcus daniellii* Benth. Eur. J. Biochem. *31*, 221-225.

VAN DER WEL, H., and LOEVE, K. 1973. Characterization of the sweet-tasting protein from *Dioscoreophyllum cumminsii* (Stapf) Diels. FEBS Lett. 29, 181-184.

WEINER, A. M., PLATT, J., and WEBER, K. 1972. Amino-terminal sequence analysis of proteins purified on a nanomole scale by gel electrophoresis. J. Biol. Chem. *247*, 3242-3251.

CHAPTER 18

Frank R. Dastoli
Robert J. Harvey

Miracle Fruit Concentrate

A native shrub of West Africa, *Synsepalum dulcificum*, yields a small red berry called miracle fruit which contains a glycoprotein that has no appreciable taste of its own, but causes sour foods to taste sweet and improves the flavor of many foods.

In its classification of the Foods of Africa (Jardin 1970), Food and Agriculture Organization (FAO) includes the fruit of *Synsepalum dulcificum* as a sweetener which is eaten in moderate quantities. Recent studies in tropical West Africa confirm the fact that this fruit is still quite commonly eaten, particularly by children in villages located in areas where the tree grows.

The first adequate description of *Synsepalum dulcificum* and its unique properties was given in 1852 by Daniell, a British surgeon stationed in West Africa (Daniell 1852). However, it was not until 1964 that the first attempts were made to isolate the active principle from the berries (Inglett *et al.* 1964). International Minerals and Chemical Corporation of Libertyville, Illinois, intrigued by the sweet-inducing properties of miracle fruit, initiated a project to identify the active principle with a view to its subsequent chemical synthesis. This work revealed that the active principle is labile, insoluble in ordinary aqueous and organic solvents, and of proteinaceous nature. In light of the incompatability of the objectives of the project and the proteinaceous nature of the active principle, the project was discontinued.

In 1968, two different groups, Kurihara and Beidler (Florida State University) and Brouwer, van der Wel, Francke and Henning (Unilever Research Laboratory, The Netherlands) arrived independently at the isolation and partial characterization of the active principle in miracle fruit. Agreement between the results of the two groups was very good, differing only in the description of the sugar groups contained in the protein. In summary, the active principle is a glycoprotein, with a molecular weight of 44,000 to 48,000. It is of a basic nature, with sugar groups of xylose and arabinose found by Kurihara and Beidler (1968) and glucose, ribose, arabinose, galactose and rhamnose as characterized by Brouwer *et al.* (1968). Considerably more basic study is required to establish the exact composition and configuration of the protein and its mechanism of action. Such studies are under way at the Miralin Corporation, Hudson, Massachusetts.

Two theories have been proposed to explain the miracle fruit phenomenon. Dzendolet (1969) suggested that anions of some acids, for example, the citrate ion of citric acid, are sweet but are normally inhibited by the sour taste. Miracle fruit, by blocking the sour receptor sites, would allow the sweet taste of the anion to be perceived. Kurihara and Beidler (1969) suggest that the glycoprotein binds to the receptor membrane near the sweet receptor site so that it will fit the sugar groups attached to the glycoprotein, producing a sweet taste. Neither of these theories is adequate to explain the fascinating but complex taste-modifying effect of miracle fruit.

Since equally sour acids are not equally sweet after miracle fruit, sourness (or pH) alone cannot be the property of acids that cause sweetness after miracle fruit. Most likely pH is but one part of the overall mechanism responsible for conformational change, pattern of electric charge on the protein for site binding, and regulation of the extent to which the active principle protein interacts with any substance. The increased effect with organic acids, particularly citric, over inorganic acids like hydrochloric, may suggest an affinity for particular groups on the acid, and thus a "fit" or stereochemical interaction. This effect, coupled with an association with the sweet-receptor site, seems to indicate the need for a rather complex series of events to occur for the sweet-inducing phenomenon of miracle fruit to take place. Studies designed to test the role of each parameter to the exclusion of others are essential to correct delineation of the phenomenon. Such studies are also under way at Miralin.

The botanical plant species of miracle fruit is *Synsepalum dulcificum* (Schum.) Daniell, Sapotaceae. It is indigenous to tropical West Africa, from Ghana to the Congo. The red fruit grows on a shrub-like, multibranched tree with dense foliage. In its native habitat, the bush grows very slowly, attaining a height of only 4 ft after 10 yr. In tropical West Africa, it bears fruit in December and again in June. A young tree begins to bear fruit when it is about 3 to 5 yr old. Trees estimated to be around 40 yr old still bear fruit. A tropical climate is required to grow the trees.

The fruit berries are ellipsoidal, about 0.75 in long and composed of a thin layer of pulp surrounding a single large seed shaped like an olive. In the fresh fruit the taste-modifying activity decays very rapidly after picking. Freezing the berry immediately after picking results in retention of the activity for a considerable period; however, the activity diminishes very rapidly upon thawing (See Fig. 18.1).

CLOSE-UP OF BERRIES ON TREES

SYNSEPALUM DULCIFICUM TREE   MIRACLE FRUIT BERRIES

FIG. 18.1. THE MIRACLE FRUIT TREE AND BERRIES AT PEAK HARVEST
CONDITION

The effectiveness of miracle fruit can best be appreciated by the statements of a pioneer investigator in the field (Inglett *et al.* 1964), "The quality of the miracle fruit-induced sweetness is more desirable than any of the known natural or synthetic sweeteners. The natural flavor of citrus fruits eaten after miracle fruit, for example, is vastly superior to any sugar-sweetened citrus product. Many of the delicate flavors, which would ordinarily be masked by sugar, appear to be well blended with the sweetness. The delightful flavor of fresh strawberries eaten after miracle fruit is so wonderful that it defies adequate description."

Obviously, such a substance should have tremendous commercial potential and the question can rightfully be asked, "Why hasn't it been commercialized until now?" The primary reason has to do with the very labile nature of the active principle in the fruit and the facts that: (1) the active principle defies synthesis because of its complex proteinaceous nature; and (2) the horticultural technology required to cultivate successfully the millions of trees needed to supply significant quantities of the protein was not developed until quite recently.

The Miralin Corporation, through horticultural research, has succeeded in producing hybrid plants that provide several times the annual yield of the average heterogeneous trees. Development of identical clone stock material now makes it economically feasible to plant orchards of several million trees which will be capable of supplying commercial quantities of Miracle Fruit Concentrate (MFC). Biochemical research at the Miralin Corporation has also resulted in the development of an economical process for the preparation of a stable form of MFC.

One of the most obvious applications of MFC is as an essentially noncaloric sweetener and flavor enhancer. A new, safe, noncaloric sweetener with excellent taste properties obviously is in great demand. As pointed out previously, miracle fruit appears to be entirely safe for human consumption. FAO has included it on its list of Foods of Africa (Jardin 1970). There have been no adverse reports associated with its long-term human consumption in Africa nor from acute oral toxicity tests in hamsters, mice, and rats. The nutrient composition of MFC is shown in Table 18.1.

The Miralin Corporation has either conducted or supported a series of feeding studies in several animal species, so as to corroborate the metabolic safety or MFC evidenced by its long history of safe use by man. The LD_{50} tests on mice, by the Mason Research Institute of Worcester, Massachusetts, resulted in the conclusion that no actual

TABLE 18.1

MIRALIN MFC COMPOSITION

Nutrient	Percentage
Protein	1.90
Hexose	0.10
Crude fiber	0.40
Fat	0.20
Morex*	69.30
Ash	18.50
Moisture	9.60
Total constituents	100.0%

*Hydrolyzed cereal solids added to improve spray drying and tabletting.

LD_{50} could be determined, since no mice died at the maximum achievable single-dose rate of 5 gm/kg body weight. An acute oral toxicity test was conducted at the U.S. Army Natick Laboratories, Natick, Massachusetts, in hamsters. Three groups of animals were fed MFC with a maximum dose level 3,000 times human consumption. No effect, versus the controls, was observed in any pathological, physiological, biochemical and cytological parameter measured. A chronic feeding study was undertaken in 1971 in rats by the Food & Drug Research Laboratories of Maspeth, Long Island, New York. As of this date, four generations of rats have received continuous feeding of MFC with no significant differences seen between experimentals and controls. The results of these studies are to be published shortly. This research is continuing, and includes long-term safety in rats and dogs, teratology in rats and rabbits, as well as metabolic pathway tracing. Three distinct process improvements have been developed since the beginning of the Miralin Corporation's work on MFC, and such material has continued in new feeding studies.

Miracle Fruit Concentrate, in its stable dry form, can be used with food products in either of two ways: a separate unit application (SUA) taken prior to eating the food to be sweetened, or incorporated into certain food products in much the same way as other sweeteners are used. The separate unit application, that is, an amount of MFC tailored to last anywhere from 10 min to several hours, can be incorporated into a tablet or drop form. This Miracle Fruit Drop (MFD) is chewed prior to eating an unsweetened meal or snack just as one would eat a pressed breath mint or boiled candy. Thereafter, food substances containing natural organic acids will taste sweet, as if sweetened with sugar. MFC can also be used by direct incorporation application (DIA) in such foods as chewing gum and hard candies, and certain other dessert and snack products. The sweetness imparted to these items is derived from incorporated food acidulants released after the tongue has been exposed to MFC contained in the food product. However, MFC has unique properties that no other sweetener or flavor enhancer has. To use MFC as other sweeteners have been used in the past would be to overlook the real potential of miracle fruit.

A series of dietary programs are under development at Miralin based upon the use of a tablet or drop which is taken before eating. If one then eats a meal or snack which is either prepared in accordance with the carefully tested Miralin menus and recipes, or consumes Miralin's special dietary food products, one can experience a completely satisfying flavor and sweetness, and yet substantially reduce calorie and salt intake. It is possible for a person with special dietary needs to follow such specific programs as a low-salt diet, a low-fat diet, or just a low-calorie diet, while still enjoying excellent flavor and sweetness.

Miralin Corporation has developed a line of low-calorie, low-sodium and low-sugar foods. They are designed to be used with the Miracle Fruit Drop to produce a sweetness and flavor comparable to sugar-containing products and with from 6 to 27% of the usual calories of the same foods formulated with sugar. Some typical initial products, which should be introduced in select markets before the end of 1973, are shown in Table 18.2. Other products are in the development stage and will be added to the line so as to offer a wide variety of dietetic products.

TABLE 18.2

CALORIC CONTENT OF SOME MIRALIN PRODUCTS VERSUS TYPICAL SUGAR SWEETENED COUNTERPARTS

Product	Serving Size	Miralin Product (Cals/serving)	Sugar Sweetened Counterpart * (Cals/serving)	Calorie Reduction %
Ice tea w/lemon	8 oz.	6	35	83
Orange gelatin	4 oz	8	80	90
Strawberry gelatin	4 oz	8	80	90
Lemon fruit drink	8 oz	6	85-100	93
Grape jelly	1.25 oz	16	125	87
Apple jelly	1.25 oz	25	125	80
Berry table syrup	1.25 oz	18	125	86
Salad dressing	1.25 oz	14	138	90
Barbeque sauce	1.25 oz	16	60	73
Miralin MFD	One Drop	2		

*Close approximations of leading brand products.

Foods not normally sour are not sweetened by miracle fruit, and so, meats, bread and similar foods are not sweetened. The flavor of vegetables seems to be improved by miracle fruit. This may be due to the addition of a very weak sweet taste, or a reduced response to the acids normally present in all vegetables. Test subjects report that vegetables taste as though they have been moderately or adequately salted when eaten in conjunction with MFC. A diet plan book containing particular tasty Miralin recipes and menus of low calorie content also is available from the Miralin Corporation. Comprehensive nutritional and calorie information is presented for each recipe. The book will be continually enriched by the addition of new recipes as they are developed in the Miralin kitchens and consumer-tested.

SUMMARY

Fruit of the miracle fruit tree contains a glycoprotein with no apparent taste of its own but the unique ability to make sour substances taste sweet and to improve the flavor of many foods. Miracle Fruit Concentrate (MFC), a stable product form of the miracle fruit developed by the Miralin Corporation has the flexibility of being used in either of two ways: taken separately prior to eating unsweetened foods or added directly to foods.

Through the former approach, that is, a separate unit application before eating a food or snack, it is possible to introduce a completely new and satisfying dietary program which meets most of the needs of contemporary society—needs quite different from those of earlier periods during which many of our present dietary practices were acquired. The Miralin Corporation introduced such a program in 1973. It has also developed the horticultural technology to grow large numbers of trees for production of sufficient quantities of MFC, the biochemical procedures for a stable, high-yield MFC process, and a line of low-calorie, no-sugar, low-sodium dietary food products.

BIBLIOGRAPHY

BROUWER, J. N., VAN DER WEL, H., FRANCKE, A., and HENNING, G. T. 1968. Miraculin, the sweetness-inducing protein from miracle fruit. Nature 220, 373-374.

DANIELL, W. F. 1852. On the *Synsepalum dulcificum*, De Cand.; or, miraculous berry of Western Africa. Pharm. J. *11*, 445-448.

DZENDOLET, E. 1969. Theory for the mechanism of action of "miracle fruit". 1969, Percept. and Psychophys. *6*, 187-188.

INGLETT, G. E., DOWLING, B., ALBRECHT, J., and HOGLAN, F. 1964. A new concept in sweetness—taste modifying properties of miracle fruit (*Synsepalum dulcificum*). Presentation at the 148th National American Chemical Society Meeting, Chicago, Illinois, Aug. 31-Sept. 4.

INGLETT, G. E., DOWLING, B., ALBRECHT, J., and HOGLAN, F. 1965. Taste-modifying properties of miracle fruit (*Synsepalum dulcificum*) J. Agr. Food Chem. *13*, 284-287.

JARDIN, C. 1970. List of Foods used in Africa, 2nd. Ed., U.S. Dept. HEW., Bethesda, Md. U.S.A. p. 23.

KURIHARA, K., and BEIDLER, L. M. 1968. Taste-modifying protein from miracle fruit. Science *161*, 1241-1243.

KURIHARA, K., and BEIDLER, L. M. 1969. Mechanism of the action of taste-modifying protein. Nature *222*, 1176-1179.

CHAPTER 19

Marvin K. Cook
B. Harry Gominger

Glycyrrhizin

Licorice, well-known for centuries and widely used, is obtained from the roots of *Glycyrrhiza glabra*, a small shrub, grown and hand-harvested in Europe and Central Asia. The licorice root reaches maturity in 4 yr and despite its abrupt removal from the earth, the fine root network remaining in the soil, self-propagates. The roots contain from 6 to 14% glycyrrhizin; the balance comprises gums, resins, asparagine, sugars, starches, and color bodies.

Licorice is the only botanical possessing significant amounts of glycyrrhizin which occurs in nature as the mixed calcium and potassium salts of glycyrrhizic acid. Spanish root is considered the sweetest, while the Near East products tend to be more bitter, concommitant with the increase in glycyrrhizin content. Glycyrrhizin is a triterpenoid glycoside of the β-amyrin series. It contains a sugar moiety having two glucuronic acid units (Lythgoe and Trippett 1950). The structural aspects of glycyrrhizin are discussed in Chapter 20.

Modern licorice technology, while covered rather superficially in the literature (Nieman 1957; Cook 1970, 1971), still remains a trade secret. The world's leading manufacturer, MacAndrews and Forbes Company, uses an advanced countercurrent extraction process in which the most concentrated liquor passes over fresh root while fresh water scrubs the nearly spent root. Fundamentally, the manufacturing procedure consists of the careful preblending of roots from various foreign sources, shredding, extraction with hot water, followed by a concentration and drying of the aqueous extracts.

This procedure yields a variety of licorice products, including block licorice, licorice powder, liquid extracts and glycyrrhizin. Ammoniated glycyrrhizin is manufactured by a special process in which only the glycyrrhizin is removed from the total extract (MacAndrews and Forbes Co. 1970). The producer has developed a number of useful by-products from exhausted licorice root, including paperboard, agricultural fertilizers, garden mulch, and pigment dispersants.

The extracts are universally employed in the flavoring and sweetening of pipe, cigarette and chewing tobaccos. They function as natural sweeteners, basic flavorants, and humectants. Tobacco products treated with licorice extracts show a marked reduction in harshness, better moisture retention and improved burning qualities (Cook 1967). Licorice extracts are regularly used in confectionery manufacture. In the United States, confections are generally prepared with licorice extracts in combination with anethol. The European market favors natural licorice exclusively because of its unquestioned superiority in strength and quality. Other applications for refined licorice are found in pharmaceutical flavoring (Cook 1955) where it is routinely employed for repressing the unpalatibility of many oral medicaments. Some segments of the flavor industry have long utilized these extracts (Cook 1971) in root beer, chocolate, vanilla, liqueur, and other flavors. Ammonium glycyrrhizin (AG), the fully ammoniated salt of glycyrrhizic acid, is commercially available as a high-purity, spray-dried, brown powder. Further treatment and repeated cyrstallizations yields the more costly, colorless salt, monoammonium glycyrrhizinate (MAG). Both derivatives have the same degree of sweetness but they differ markedly from each other in solubility properties and sensitivity to pH.

AG is the sweetest substance on the FDA list of natural GRAS flavors. It has long been known that AG is 50 times sweeter than sucrose. In the presence of sucrose, AG is around 100 times sweeter than sucrose alone. This synergistic effect provided the basis for a U.S. patent, issued in 1966 (Muller 1966). The patent teaches that if 1 lb AG is combined with 100 lb sucrose, the sweetness of this mixture is equivalent to 200 lb cane sugar.

In 1967, MacAndrews and Forbes Company obtained a second patent relating to the application of AG in potentiating the flavor of cocoa and other cacao products (Morris 1967). Our more recent investigations have demonstrated that the natural cocoa flavor is intensified to a point where a 25% reduction in cocoa content can be effected without changing acceptability. This is a greater reduction of cocoa than is shown in the patent.

AG, a relatively stable compound, is completely soluble in hot and cold water, propylene glycol, and hydroalcoholic solutions. While AG has no discrete thermal decomposition point, it can be subjected to temperatures up to 105°C for brief periods. At a pH of approximately 4.5 and lower, AG tends to precipitate; this characteristic generally precludes its use in most carbonated beverages, except the low-acid types such as root beer, chocolate, and cream soda.

AG has served for a very long time as a foaming agent and flavor enhancer in root beer. Low-calorie carbonated beverages can easily be prepared with AG. A typical formula comprises 5% cane sugar and 0.07% AG, based on the finished beverage. This drink is equivalent in sweetness to a regular 12% sugar-based root beer and provides better than a 50% reduction in calories.

Because of its complete solubility in propylene glycol and its ability to improve dramatically certain flavorants, AG is gaining wider recognition as a useful adjunct in the arsenal of products for the flavor industry. Particularly good results in maple, root beer, rum, walnut, butterscotch, chocolate, honey, and pickle-spice compositions have been achieved in commercial practice (Mac-Andrews and Forbes Co. 1971).

A number of manufacturers have reported that they have successfully replaced saccharin with a modified AG in sugar-free chewing gum. AG appears particularly impressive in the augmentation of spearmint and peppermint flavors. Inexplicably, it frequently induces an untoward taste effect in some citrus-flavored products.

Apart from its usefulness as a masking and debittering agent in unpleasant-tasting medicines, the pharmaceutical industry has now recognized the value of AG in other products, including mouthwashes, aerosol breath deodorants, chewable vitamins, and aspirin. In mouthwashes, AG serves as a foaming agent, flavorant, flavor potentiator, and natural sweetener. These multiple attributes are not offered by any other single compound, either natural or synthetic.

It has come to our attention that the Japanese food industry has incorporated licorice extract and more recently glycyrrhizin salts in hydrolyzed vegetable proteins, soy sauce and bean paste in order to modify the characteristic saltiness of these products. We have been informed that ammoniated glycyrrhizin is effective in augmenting favorably the flavor of imitation meat products derived from the controlled pyrolysis of amino acids and sugars.

The manifold applications of AG are far too numerous to detail here. However, it should be mentioned that particularly noteworthy results have been reported to us recently by manufacturers of sugar substitutes, vegetable toppings, dehydrated soups, and various low-calorie products. As a general rule, the levels of AG in actual use are from 30 to 500 ppm and sometimes distinctly less in the presence of sugar.

It has been our observation that the flavoring effects and intense potency of AG are not always clearly appreciated. The final taste in any given product is obviously dependent upon many factors, including the amount of AG utilized, type of flavors employed with it, pH, natural sugar content, and the basic flavor of the product itself. Considerable effort is frequently required to arrive at the desired optimum taste, and in this respect AG appears much more critical and demanding than most other flavorants.

Comparative taste panel studies with AG-treated products may lead to confusing results because of the subtle, long-lasting properties in the mouth. We are of the opinion that a protein-bonding may take place with the oral mucosa which conceivably could account for the protracted persistence or recurrence of the flavor. Consequently, in taste-panel studies, the controls may easily be potentiated because of the flavor carry-over. This seems particularly true when attempting to establish relative sweetness levels with sugars. Repeated cross-testing of the control and the test samples tends more often than not to confuse the tester.

For a long time, we have been aware of the lingering licorice taste imparted by the root derivatives which some find highly undesirable. A good deal of practical research has gone into the development of products which ideally would have a repressed licorice character and simultaneously, unimpaired sweetness. Positive indications from diverse industries lead us to believe that our most recent efforts in this direction have been fruitful. At this time, we are not free to disclose the nature of these innovations and can only say that they comprise GRAS compounds and are commercially available in production quantities.

Finally, we come to monoammonium glycyrrhizinate or MAG, a most important pure white salt that demonstrates pH insensitivity over a wide range. MAG has very poor solubility in water and alcohol. However, after special processing, MAG readily dissolves in water, causing increased viscosity and slight opacification of the liquid.

Many dentifrices are sweetened exclusively with saccharin. They also contain foaming agents such as sodium lauryl sulfate and hydrocolloid thickeners and stabilizers (Cook 1958). Because of its sweetening and nonfermentative properties, together with its excellent foaming, viscosity-building, and flavor-reinforcing action, MAG is ideally suited for dentifrice manufacture.

MAG appears very effective in rounding out the harsh notes so often encountered in spice-mint blends which are regularly applied in flavoring toothpastes and mouthwashes. Furthermore, MAG, unlike AG, does not contribute any color to these products. The pH of aqueous MAG solutions is around 4.5.

Food applications with MAG have obviously been broadened so that this root derivative can easily be applied to high-acid preparations such as beverages and gelatin desserts. We also envision use in salad dressings, viable and canned fruits, fermented beverages, and acidic confections. There is probably not enough licorice root grown and processed to meet all the requirements of industry, particularly in the replacement of saccharin.

New applications are constantly being found for AG and MAG in many industries. The surprising fact is that it took so long for their valuable properties to be recognized and applied.

BIBLIOGRAPHY

COOK, M. K. 1955. Pharmaceutical flavoring. Drug Cosmetic Ind. *76*, No. 5, 624-625.

COOK, M. K. 1958. Modern toothpaste manufacture. Drug Cosmetic Ind. *82*, No. 3, 314-316.

COOK, M. K. 1967. Aromatized tobaccos. U.S. Pat. 3,342,186. Sept. 19.

COOK, M. K. 1970. Ammoniated glycyrrhizin, a useful natural sweetener and flavor potentiator. Flavour Ind. *1*, No. 12, 831-832.

COOK, M. K. 1971. Flavour creation. Flavour Ind. *2*, No. 3, 155-156.

COOK, M. K. 1971A. Licorice root derivatives for use in cosmetic products. Drug Cosmetic Ind. *109*, No. 10, 50, 52, 138-140.

LYTHGOE, B., and TRIPPETT, S. 1950. The constitution of the disaccharide of glycyrrhetinic acid. J. Chem. Soc. 1983-1990.

MACANDREWS & FORBES COMPANY. 1970. Personal communication. Camden, N.J.

MACANDREWS & FORBES COMPANY. 1971. Ammoniated glycyrrhizin, natural sweetener and flavor potentiator. (Bulletin). Camden, N.J.

MORRIS, J. 1967. Potentiation of chocolate flavor with ammonium glycyrrhizin. U.S. Pat. 3,356,505. Dec. 5.

MULLER, R. E. 1966. Sucrose-ammoniated glycyrrhizin sweetening agent. U.S. Pat. 3,282,706. Nov. 1.

NIEMAN, C. 1957. Licorice. Advances in Food Research, Vol. 7. Academic Press, New York.

CHAPTER 20

J. E. Hodge
G. E. Inglett

Structural Aspects of Glycosidic Sweeteners Containing (1'→2) - Linked Disaccharides

Present interest in finding a low-calorie, innocuous sweetener to replace those of doubtful healthfulness has prompted this review of the properties and structures of sweet glycosides of natural origin that are free of nitrogen and sulfur. Stevioside, osladin, and glycyrrhizin are nonaromatic, natural glycosides that are intensely sweet. Stevioside, extracted from the leaves of a Paraguayan shrub, is 280 to 300 times sweeter than sucrose and nearly as sweet as saccharin. Osladin is the recently identified sweet principle of Polypody fern rhizomes; it is said to have a saccharin-like sweetness. Glycyrrhizin has long been extracted from licorice roots for use as flavoring in pharmaceuticals and confections. Ammoniated glycyrrhizin, in its commercial form, is rated about 50 times sweeter than sucrose; however, its relative sweetness is difficult to assess because of accompanying licorice flavors.

Glycosidic sweeteners are also prepared by chemical conversion of the naturally occurring citrus flavonoids, naringin and neohesperidin, to the dihydrochalcones. Naringin dihydrochalcone is 0.4 as sweet as saccharin, whereas neohesperidin dihydrochalcone is 7 times as sweet. Although the dihydrochalcones contain aromatic groups, ingested flavonoids are generally harmless, and rat-feeding tests on the dihydrochalcones have revealed no harmful effects (Horowitz and Gentili 1971).

All five of these intensely sweet glycosides contain a strongly hydrophobic aglycone, a (1'→2)-linked disaccharide, and another important hydrophilic function attached to the aglycone but spaced apart from the disaccharide. They all show roughly similar molecular structures and shapes. The flavor properties and established constitution of each of these glycosides are reviewed before their common structural relationships are discussed.

STEVIOSIDE

The sweet herb of Paraguay (Yerba dulce), called variously Caa-ehe, Azuca-caa, Kaa-he-e, and Ca-a-yupe by the Guarani, has long been the source of an intense sweetener. Natives use the leaves of this small shrub to sweeten their bitter drinks. The plant was first given the botanical name *Eupatorium rebaudianum*, but this was later changed to *Stevia rebaudiana* Bertoni; recently, the name *Stevia rebaudiana* (Bert.) Hemsl has appeared. For the sweet, crystalline glycoside that has been extracted from the leaves of *S. rebaudiana*, the name stevioside was adopted by the Union Internationale de Chimie in 1921. Historical accounts of knowledge on stevioside and proposals for cultivation of *S. rebaudiana* for commercial use of stevioside as a sweetener have been reviewed by Bell (1954), Fletcher (1955), and Nieman (1958). Wood *et al.* (1955) reported a method for extracting stevioside in 7% yield from air-dried leaves of the Paraguayan plant.

Bridel and Lavieille (1931A) reported stevioside as a white, crystalline, hygroscopic powder, approximately 300 times sweeter than cane sugar. Very small amounts on the tongue gave a delectable sweetness, like the leaves of the plant; however, large amounts tasted sweet at first then distinctly bitter. On the other hand, no bitter taste was attributed to stevioside by Nieman. By taste panel, Pilgrim and Schutz (1959) determined a relative sweetness of 280 for stevioside as against 306 for saccharin (sucrose = 1.00).

Bridel and Lavieille (1931B) showed that stevioside is rapidly hydrolyzed by enzymatic material extracted from the vineyard snail, *Helix pomatia*, giving rise to 3 moles of D-glucose and 1 mole of a tasteless, acidic aglycone which they named steviol. Acidic hydrolysis gave the same percentage of D-glucose but a different aglycone named isosteviol. Their work on the constitution of stevioside was repeated, confirmed, and extended by Wood *et al.* (1955). The identity and positions of the sugars linked to steviol were fixed by Wood *et al.* and by Vis and Fletcher (1956). When the absolute configuration of the diterpenoid aglycone was finally resolved by Mosettig *et al.* (1963), the complete structures of steviol and stevioside were established as shown in Fig. 20.1.

The disaccharide of stevioside is sophorose (2-*O*-β-D-glucopyranosyl-β-D-glucopyranose). It is linked to the tertiary α-hydroxyl at C-13 of steviol, whereas the monosaccharide, β-D-glucopyranose, is condensed with the sterically hindered α-carboxyl at C-4. The two sugars are, therefore, appended to the same side of the rigid aglycone at opposite ends. Alkaline splitting of the unusual D-glucose-to-carboxyl linkage produced levoglucosan (1,6-anhydro-β-D-glucopyranose) and the sophoroside of steviol, which was devoid of sweetness (Wood *et al.* 1955; Vis and Fletcher 1956).

Pomaret and Lavieille (1931) reported that stevioside readily passes through human elimination channels in its original form. It did not appear to be toxic to guinea pigs, rabbits, or chickens. Furthermore, there are no recorded reports of ill effects in Paraguayan users of the leaves of *S. rebaudiana*. Nevertheless, the

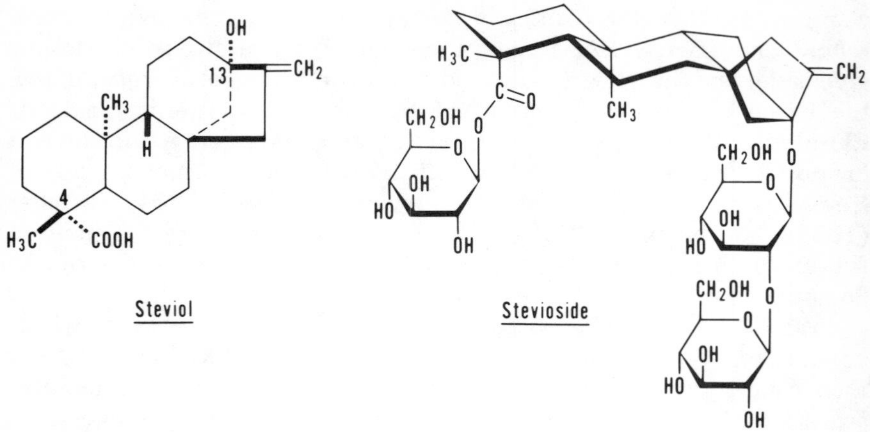

FIG. 20.1. AGLYCONE AND SWEET GLYCOSIDE OF *STEVIA REBAUDIANA*

long-term effects of ingestion of stevioside would have to be investigated carefully before it could be considered for human use as a sweetener in the United States. The diterpenoid aglycone, steviol, has shown specific physiological activity (Vignais *et al.* 1966) and weak antiandrogenic effects (Dorfman and Nes 1960). It remains to be proved that stevioside does not split to form *any* steviol in the human digestive tract.

OSLADIN

The sweet taste of rhizomes of the widely distributed fern, *Polypodium vulgare* L., has attracted the interest of many chemists and pharmacists. Van der Vijver and Uffelie (1966) and others have shown that the sweet substance is not glycyrrhizin, as was once proposed. Many constituents of the rhizomes have been isolated, but the substance that resembles saccharin in sweetness was isolated and identified only recently (Jizba and Herout 1967, Jizba *et al.* 1971A). The name osladin is based on the Czech name for polypody, osladic. Osladin comprised only 0.03% of the dry weight of the rhizomes. Its chemical structure, shown in Fig. 20.2, was revealed as a bis-glycoside of a new type of steriodal saponin.

The glycoside that results by replacement of the monosaccharide radical with hydrogen in the formula of Fig. 20.2 was isolated separately and named polypodosaponin. Its absolute configuration was determined by Jizba *et al.* (1971B). They made no comment on its taste.

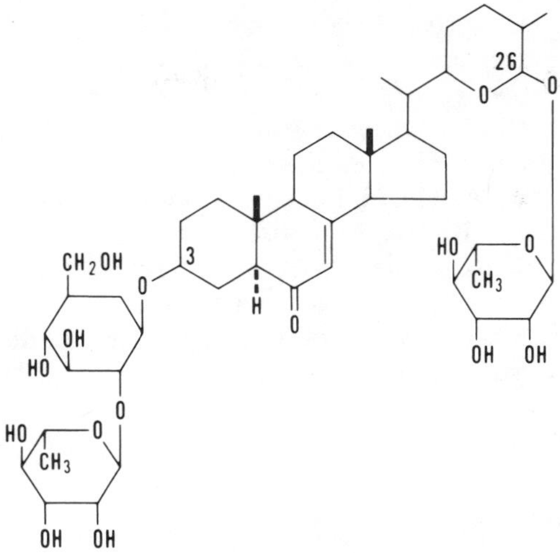

FIG. 20.2. STRUCTURE OF OSLADIN, SWEET PRINCIPLE OF *POLYPODIUM VULGARE* L. RHIZOMES

The disaccharide of osladin was shown to be neohesperidose, 2-*O*-α-L-rhamnopyranosyl-β-D-glucopyranose. The glycosidic linkage was shown to be beta by cleavage of neohesperidin, with a specific β-glucosidase from *Aspergillus wentii*. Therefore, neohesperidose is in the same glycosidic configuration in osladin as is found in the intensely sweet neohesperidin dihydrochalcone. Neohesperidose is only slightly sweet, if at all (Koeppen 1968, Horowitz and Gentili 1969). The configuration of the glycosidic linkage of the monosaccharide, L-rhamnose, has not been determined. It is probably α-L-, as shown, because this corresponds to the β-D-linkage usually found in natural glycosides. The molecular structure of osladin resembles that of stevioside; it shows a (1'→2)-linked disaccharide at one end of the aglycone and a monosaccharide at the other.

The C-26-*O*-methyl polypodosaponin shows hemolytic activity and a strong inhibition of fungal growth (Jizba *et al.* 1971B). Even without regard to biological activity, the very low concentration of osladin found in polypody rhizomes dims the prospects for developing osladin as a commercial sweetener.

GLYCYRRHIZIN

Glycyrrhizin and glycyrrhizic acid are trivial names for the main glycosidic constituent and sweet principle of licorice root (*Glycyr-*

rhiza glabra L.). Chapter 19 contains a discussion of the technological aspects of ammoniated glycyrrhizin, the article of commerce,
and its useful properties as a sweetener and sweetness enhancer. Only
the investigations that have established the structure of glycyrrhizin
are presented here.

Glycyrrhizic acid exists in licorice root as the calcium-potassium
salt in association with other constituents, such as starch, gums,
sugars, proteins, asparagine, flavonoids, and resins. Although this
glycoside is difficult to free of nitrogen, minerals, color, and licorice
flavor, it has been isolated in pure form. The established chemical
structure of glycyrrhizic acid is shown in Fig. 20.3. It is a glycoside

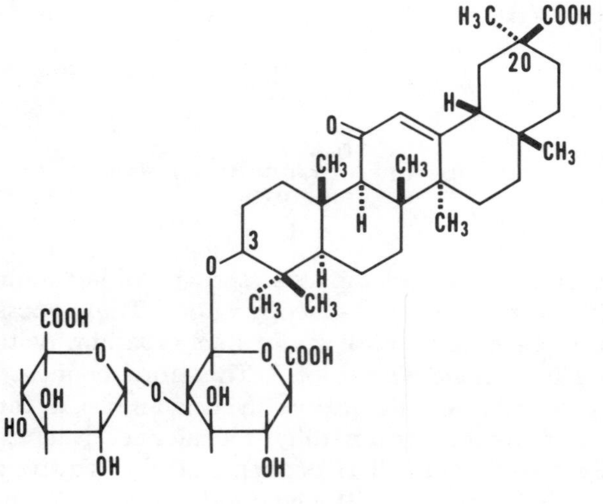

FIG. 20.3 STRUCTURE OF GLYCYRRHIZIC ACID,
SWEET PRINCIPLE OF LICORICE ROOT, *GLYCYR-
RHIZA GLABRA* L.

of the triterpene, glycyrrhetic acid, which is condensed with
O-β-D-glucuronosyl-(1′ → 2)-β-D-glucuronic acid. Colorless, crystalline
glycyrrhizic acid was first isolated by Tschirch and Cederberg (1907).
Although the empirical formula and optical inactivity that they
reported were incorrect, the other properties given and their isolation
procedures have been repeatedly reproduced (Tschirch and Gauchmann 1908, Lythgoe and Trippett 1950, Marsh and Levvy 1956,
Vovan and Dumazert 1970A). A "very sweet" taste was ascribed to
the free, tribasic acid. It was practically insoluble in cold water and
soluble in hot water. The free acid and its ammonium and potassium

salts formed pasty gels upon cooling the warm aqueous solutions; however, they could be recrystallized from glacial acetic acid or alcohol. An "intensely sweet" taste was ascribed to the water-soluble salts. Tschirch and Cederberg hydrolyzed potassium glycyrrhizinate in dilute sulfuric acid to obtain the crystalline aglycone, glycyrrhetic acid. It was insoluble in water and tasteless. Glucuronic acid was identified in the mother liquor by Tschirch and Gauchmann (1908).

Voss et al. (1937A) first obtained the correct empirical formula for glycyrrhizin, and Voss and Pfirschke (1937B) showed that the glucuronic acid found is linked to the hydroxyl group of glycyrrhetic acid as a disaccharide. Lythgoe and Trippett (1950) proved the $(1' \rightarrow 2)$-interglycosidic linkage, also the pyranoside rings of the di-glucuronic acid moiety, by periodate oxidation and methylation analysis. From optical rotation data, they could agree with Voss and Pfirschke's supposition that the interglycosidic linkage is beta. Both the interglycosidic linkage and the linkage to the aglycone were shown to be beta by Marsh and Levvy (1956). They hydrolyzed pure ammonium glycyrrhizinate with a β-glucuronidase isolated from mouse liver. This enzyme was shown to activate only β-glucuronides in pyranoside ring-form. Vovan and Dumazert (1970B) have confirmed that the only sugar present in pure, crystalline glycyrrhizic acid (1970A) is glucuronic acid. Voss and Butter (1937C) showed that the C-20 carboxyl group is not sterically hindered. All three carboxyl groups are readily methylated (Brieskorn and Sax 1970).

The absolute configuration of the aglycone, glycyrrhetic acid, is determined as the result of investigations too numerous to cite completely. Important contributions were made by Ruzicka et al. (1936, 1937A, 1937B, 1939, 1943), Voss et al. (1937A, 1937C), and Beaton and Spring (1955). Although two isomers (18-α and 18-β)

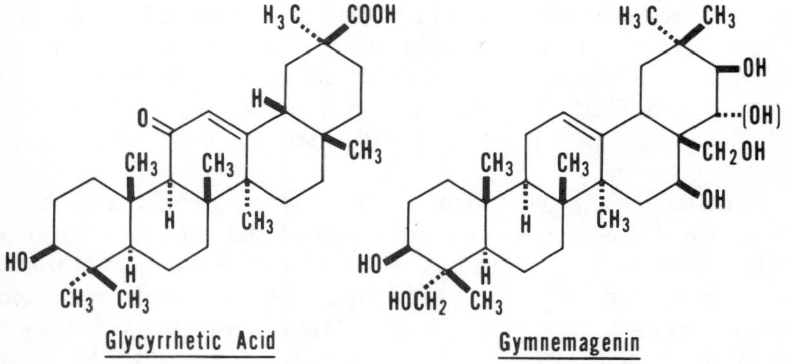

Glycyrrhetic Acid   Gymnemagenin

FIG. 20.4. STRUCTURE OF SIMILAR AGLYCONES FROM GLYCYRRHIZIN AND GYMNEMIC ACID A_1

have been isolated, Beaton and Spring have indicated that only the 18-β-glycyrrhetic acid (shown in Figs. 20.3 and 20.4) is the natural isomer that occurs in glycyrrhizin.

The aglycone of glycyrrhizin is closely related to β-amyrin, a triterpene; however, it gives medicinal effects similar to those of deoxycorticosterone acetate. Both glycyrrhizin and glycyrrhetic acid, and their respective methyl esters, show a low hemolytic activity in comparison with some other triterpenoid saponins (Schloesser and Wulff 1969).

Glycyrrhizin does not contain a second sugar radical as stevioside and osladin do. Still, the nonhindered carboxyl group at C-20 and the polar disaccharide at C-3 of glycyrrhetic acid provide for hydrophilic structures at both ends of the hydrophobic aglycone.

GYMNEMIC ACIDS

Gymnemic acids are acidic glycosides found in the leaves and roots of the tropical plant, *Gymnema sylvestre* R. Br. (Asclepiadaceae). A tree climber, this plant is found in Central and Western India, in tropical Africa, and in Australia. Chewing the leaves or roots of this plant blocks perception of the sweet taste of sugars, glycerol, saccharin, and cyclamates for hours. Stoecklin (1969A) has reviewed the history of observations on these phenomena, as well as the structural and physiological investigations of the active principles. Bartoshuk *et al.* (1969) showed that only sweet taste, and not bitter, salty, or sour, is blocked by gymnemic acids.

Sinsheimer *et al.* (1970) have shown that at least 9 closely related, acidic glycosides can be isolated from *G. sylvestre* leaves. Two of the major constituents yield glucuronic acid on hydrolysis, whereas two others (with unproved antisweet activity) yield glucose in addition to glucuronic acid. From Stoecklin's (1969A) correlation, it appears that the gymnemic acids designated A_1 and A_2, which yield glucuronic acid only, are the most effective taste-blocking compounds. For the aglycone of these two gymnemic acids, Stoecklin (1969B) has proposed the β-amyrin-like structure shown in Fig. 20.4 for gymnemagenin.

The pentacyclic, hexahydroxy triterpene is esterified with organic acids of low molecular weight and glycosylated with glucuronic acid at unknown positions. Kurihara *et al.* (1969) have shown that the genin of gymnemic acid A_1 is esterified with 1 mole of acetic acid, 2 moles of isovaleric acid, and 1 mole of tiglic acid. By hydrolysis of 1 mole of acetic acid in the genin, A_1 was converted into A_2 and the antisweet activity fell to 20% of that of A_1. Hydrolysis of all esterified groups in the genin converted A_1 into A_3, which had no antisweet activity. Gymnemic acid A_1 probably is the most active taste-blocking agent of the gymnemic acids; it suppressed the sweet taste of sodium cyclamate, D-amino acids, beryllium dichloride, and lead diacetate, but not that of chloroform.

From the two structures in Fig. 20.4, and considering that at least 4 of the 6 hydroxyls of gymnemagenin are esterified and that one or two of the others are glycosylated with glucuronic acid, it is quite possible that gymnemic acid A_1 and glycyrrhizic acid have similar molecular structures and shapes.

GLYCOSIDIC CITRUS FLAVONOID DIHYDROCHALCONES

Horowitz has presented the structural aspects of dihydrochalcone sweeteners in Chapter 16. Further information on the sensory properties and structural variations within this class of compounds is available (Inglett *et al.* 1969, Inglett 1971). It remains to be emphasized here that the naringin and neohesperidin dihydrochalcones, like the other intensely sweet glycosides previously discussed, contain a $(1' \rightarrow 2)$-linked disaccharide. Furthermore, this disaccharide, neohesperidose, is linked at one end of a hydrophobic aglycone with a hydrophilic function at the other end. Osladin presents another example of an intensely sweet glycoside of neohesperidose.

Neohesperidose is not sweet *per se* (Koeppen 1968, Horowitz and Gentili 1969); but when it is condensed with a nonsweet, hydrophobic aglycone, suitably substituted with a hydrophilic group in a specific configuration, the glycoside becomes intensely sweet. Substituting D-galactose for D-glucose in the neohesperidose of naringin dihydrochalcone does not alter its sweetness (Van Niekerk and Koeppen 1972). For the dihydrochalcones, the specificity required for the position of the second hydrophilic function (hydroxyl on phenyl ring B) is remarkable (Horowitz and Gentili 1971, Inglett 1971). A similar specificity has been demonstrated for 3'-hydroxyl and 4'-methoxyl groups on phenyl ring B of the strongly sweet isocoumarin derivative, phyllodulcin (Yamato *et al.* 1972).

STRUCTURE CORRELATIONS

The intensely sweet glycosides have several structural features in common. Three of these features are evident from the listings in Table 20.1.

Each of the 5 glycosides contains a $(1' \rightarrow 2)$-linked disaccharide; however, three different disaccharides have been identified: sophorose, neohesperidose, and the di-glucuronic acid of glycyrrhizin. The only intensely sweet glycosides which do not contain a $(1' \rightarrow 2)$-linked

TABLE 20.1

STRUCTURAL FEATURES OF INTENSELY SWEET GLYCOSIDES

Glycoside	Disaccharide	Hydrophobic Part	Polar "Probe"
Stevioside	2-O-β-D-glucosyl-β-D-glucoside	Steviol (diterpene)	β-D-glucose
Glycyrrhizin	2-O-β-D-glucuronosyl-β-D-glucuronoside	Glycyrrhetic acid (triterpene)	COOH
Osladin	2-O-α-L-rhamnosyl-β-D-glucoside	Polypodosaponin (steroid)	α-L-rhamnose
Neohesperidin dihydrochalcone	2-O-α-L-rhamnosyl-β-D-glucoside	Subst'd phloro-glucinol,-C_2H_4-	phenyl OH and OCH_3
Naringin DHC	2-O-α-L-rhamnosyl-β-D-glucoside	Subst'd phloro-glucinol,-C_2H_4-	phenyl OH

disaccharide are the synthetic β-D-glucoside and -xyloside of hesperedin dihydrochalcone discussed in Chapter 16. The β-D-glucoside can be considered as neohesperidin dihydrochalcone minus the α-L-rhamnosyl radical; it is only one-twentieth as sweet as its precursor (Horowitz and Gentili 1969, 1971).

Each of the 5 glycosides contains an extended hydrophobic aglycone, which limits its solubility in water. Removal of the disaccharides by enzymatic hydrolysis leaves a water-insoluble aglycone that is tasteless. On the other hand, removal of the monosaccharide (D-glucose) in stevioside by alkaline hydrolysis left steviol-13-sophoroside, analogous in structure to glycyrrhizic acid. Like glycyrrhizic acid, it was only very slightly soluble in cold water; but it was not sweet (Wood et al. 1955). Vis and Fletcher (1956) showed that the carboxyl group at C-4 in steviol is strongly hindered sterically; hence, even if the different spacings were not a factor, this carboxyl group would not be expected to act, presumably as does the carboxyl group in glycyrrhizic acid, to induce a sweet response.

Hydrophobe-hydrophile balance is recognized as one of several important factors for determining the intensity of sweetness of a compound. Deutsch and Hansch (1966) showed, for a series of 1-substituted-2-amino-4-nitrobenzenes, that their sweetness was a function of the partition coefficient of the compound between octanol and water. They concluded that favorable partitioning onto sweet-taste receptor sites would depend upon hydrophobic bonding, polarity of the molecule, and other factors. Systematic structural variation of other types of compounds coupled with sweetness testing has indicated the criticalness of carbon chain-length or ring-size, or molecular shape, in the hydrophobic portion of perillaldehyde oxime (perillartine) analogs (Acton et al. 1970, Unterhalt and Boeschemeyer 1971), aspartylphenylalanine ester analogs (Mazur et al. 1970; see Chapter 14), cyclamate analogs (Unterhalt and Boeschemeyer 1972), and phyllodulcin analogs (Yamato et al. 1972). However, partition coefficients between hydrophobic and hydrophilic solvents have not been reported for these series of compounds.

Simple molecules that are sweeter than sucrose, such as saccharin, cyclamates, dulcin, phyllodulcin, perillaldehyde oxime, D-tryptophan and a related indole sweetener (Hofmann 1972), aspartylphenylalanine methyl ester, and N'-acyl kynurenines, all show an important hydrophobic function related to one or more binary hydrophilic functions. Fig. 20.5 shows this division of functions for N'-formyl kynurenine, which was recently synthesized by Finley and Friedman (1973). Hydrophobic and hydrophilic

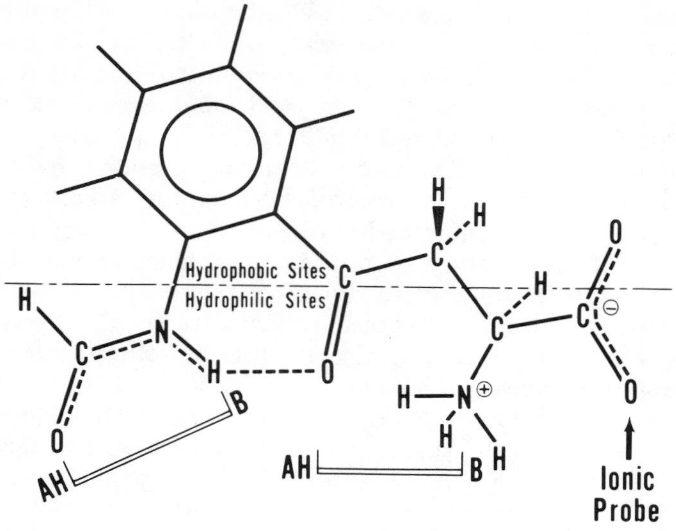

FIG. 20.5. STRUCTURE OF N'-FORMYL KYNURENINE SHOWING
HYDROPHOBIC AND HYDROPHILIC PARTS OF THE MOLECULE

binding sites on the taste buds would be occupied by hydrophobic and hydrophilic parts of the sweet-tasting compound, presumably as shown. A similar hydrophobe-hydrophile division for the larger sweet glycoside molecules is prepared below.

Another uniform feature of the 5 glycosides, not shown in Table 20.1, is the unsaturated grouping located near the middle of each hydrophobic aglycone, excepting steviol. The exo-double bond in steviol is located at the extreme end of the aglycone (Figs. 20.1 and 20.6). The aglycones of osladin and glycyrrhizin contain an α,β-unsaturated ketone group, but only the double bond is present in gymnemic acid A_1. The significance of this common structural feature is not clear; presently it seems not to be important.

Each of the 5 glycosides contains, in addition to the hydrophilic disaccharide and the hydrophobic aglycone, another hydrophilic function, designated as the polar "probe" in Table 20.1. The probe is considered to be a necessary structural feature. Numerous naturally occurring glycosides (e.g., saponins) contain terpenoid and steroid aglycones; but, because they do not contain a sufficiently polar group properly positioned to function as the probe, they are not sweet. The probe may be a sugar or phenolic radical with Shallenberger's AH,B hydrogen-bonding function (Shallenberger and Acree 1969, 1971); or it may be simply an anionic center, such as the C-20 carboxylate anion of ammonium glycyrrhizinate or the phenolate anion of naringin dihydrochalcone. The probe of neohesperidin dihydrochalcone and phyllodulcin is the 3'-OH, 4'-OCH$_3$ grouping on Ring B, which also constitutes an AH,B pair. The high stereospecificity required of the probe structure has been demonstrated and discussed for the citrus flavonoid dihydrochalcones (Krbechek *et al.* 1968, Horowitz and Gentili 1971) and for phyllodulcin (Yamato *et al.* 1972).

Still another structural feature common to the 5 intensely sweet glycosides should not be overlooked. The three-dimensional models of each (Figs. 20.6-20.9) show that they have similar gross molecular shapes. Models of the molecules form U-shaped structures, wherein the legs of the "U" are hydrophilic groups and the curved portion is the hydrophobic aglycone. In stevioside and osladin (Figs. 20.6 and 20.7), the legs of the "U" are the disaccharide and the monosaccharide. These are so positioned by the aglycone that they can approach each other very closely. A closed ring or loop would be

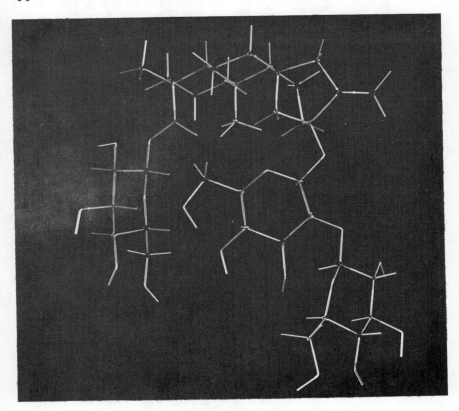

FIG. 20.6. DREIDING MODEL OF STEVIOSIDE

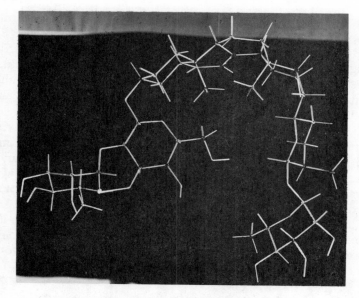

FIG. 20.7. DREIDING MODEL OF OSLADIN

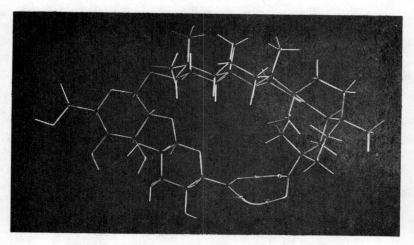

FIG. 20.8. DREIDING MODEL OF GLYCYRRHIZIC ACID

formed by hydrogen bonding between the two saccharides. In glycyrrhizin and neohesperidin dihydrochalcone (Figs. 20.8 and 20.9), loops also could be formed through hydrogen bonding—by carboxyl-to-carboxyl bonding in glycyrrhizin (Fig. 20.8) or by sugar hydroxyl to the π-bonding electronegative cloud of the phenol ring

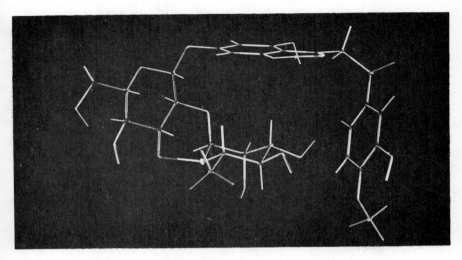

FIG. 20.9. DREIDING MODEL OF NEOHESPERIDIN DIHYDRO-CHALCONE

(Fig. 20.9). There is no experimental evidence for looping of these molecules through hydrogen bonding. This possibility is mentioned and illustrated only to emphasize that many AH,B bonding centers are grouped close together on the hydrophilic side of the molecule, in opposition to the many hydrophobic bonding centers on the other side. The molecules could, therefore, fit nicely between any opposing hydrophilic and hydrophobic binding sites of the taste bud cells.

Spatial models of the (1'→2)-linked disaccharides show the strong probability of intramolecular hydrogen bonding between the C-3-OH and the C-5' ring oxygen. The proposed H-bond is seen in Figs. 20.7, 20.9, and 20.10 as a bright spacer imbedded in a small ball of wax. The effect of this bond is to position the plane of one sugar ring nearly perpendicular to the other. The hydroxyl groups would then form a unique, relatively fixed steric arrangement that would not be duplicated in (1'→3-, 1'→4-, or 1'→6)-linked disaccharides.

To compare the spatial relationships between the sweetness-blocking gymnemic acid and the sweet glycosides, a hypothetical model was constructed using Stoecklin's formulation of gymnemagenin (Fig. 20.10; cf. Fig. 20.4). A (1'→2)-linked diglucuronic acid radical, like that in glycyrrhizin, is arbitrarily attached at the C-3-OH. This is the most likely site, according to other saponin structures. It must be remembered that the hydroxyl groups that are not condensed with glucuronic acid are esterified with low-molecular-weight fatty acids. The C-32-OH is strongly hindered sterically and

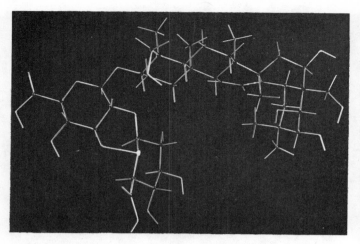

FIG. 20.10. DREIDING MODEL (HYPOTHETICAL) OF GYMNEMIC
ACID, NOT ESTERIFIED

probably could not serve as a polar probe even if it were not
esterified. The probe end of this molecule is a mass of esterified
hydroxyls and projecting methyl groups. These groups then would be
positioned over and would effectively block the most important
receptor site which requires the probe for excitation of the sweet
taste response. Gymnemic acid A_1 would occupy the secondary
receptor sites through hydrogen bonding of the glucuronic acid
groups, but it would not stimulate a sweet taste because the polar
probe is missing. Thus, it probably acts as a competitive inhibitor of
sweet-tasting compounds at the receptor site.

A schematic of the structural features of glycosides that induce an
intense sweet taste is presented in Fig. 20.11. The entire molecule
would position itself between hydrophobic and hydrophilic binding
sites of the taste bud. Because the corresponding binding sites on
molecules of the intensely sweet glycosides are numerous and
extensive, a strong and lasting sweet taste would be elicited.
Occupation of the hydrophobic and hydrophilic binding sites of the
receptor by van der Waals and hydrogen-bonding forces, respectively,
probably is only the first step in producing the sweet taste response.
The bound molecule should have a "probe" capable of bonding, by
AH,B or by ionic forces, to a more specific type of receptor site
which, by electron transfer or electrostatic charge displacement,
triggers the message of sweetness to the brain.

A similar schematic diagram (but without reference to hydro-
phobicity) has been presented by Kurihara et al. (1969) to explain

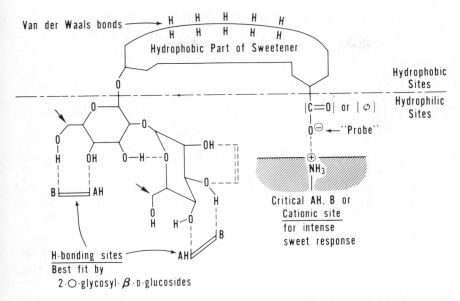

FIG. 20.11. SCHEMATIC DIAGRAM OF THE PROPOSED TYPES OF BONDING OF
SWEET GLYCOSIDES TO THE TASTE BUD RECEPTOR SITES

the action of miracle fruit glycoprotein in making sour substances taste sweet. In their hypothesis, the protein moiety occupies the secondary binding sites while the terminal saccharide moiety of the glycoprotein acts as the probe. The probe engages (bonds with) the more specific receptor site which is modified in shape or position (availability) by protons. Their hypothesis and ours are presented to stimulate further research on a very complex and mystifying problem.

What *is* the complete mechanism of sweet taste perception?

SUMMARY

The molecular structures of 5 intensely sweet glycosides, and of a sixth glycoside known to block sweet taste, are reviewed and correlated. Each sweet glycoside contains: (1) a disaccharide with an internal (1′→2)-O-glycosidic linkage; (2) an extended, hydrophobic aglycone; and (3) a second, critical hydrophilic group at the opposite end of the aglycone from the hydrophilic disaccharide. This structural correlation should promote the search for a new low-caloric sweetener to replace those of doubtful healthfulness.

BIBLIOGRAPHY

ACTON, E. M., LEAFFER, M. A., OLIVER, S. M., and STONE, H. 1970. Structure-taste relationships in oximes related to perillartine. J. Agr. Food Chem. *18*, 1061-1068.

BARTOSHUK, L. M., *et al.* 1969. Effects of *Gymnema sylvestre* and *Synsepalum dulcificum* on taste in man. *In* Olfaction and Taste, Vol. III, C. Pfaffmann (Editor). Rockefeller Univ. Press, New York, pp. 436-444.

BEATON, J. M., and SPRING, F. S. 1955. The configuration of the carboxyl group in glycyrrhetic acid. J. Chem. Soc. 3126-3129.

BELL, F. 1954. Stevioside: a unique sweetening agent. Chem. Ind. (London) 897-898 (July 17).

BRIDEL, M., and LAVIEILLE, R. 1931A. The sweet principle of the leaves of Kaa-he-e (*Stevia rebaudiana* Bertoni). Compt. Rend. *192*, 1123-1125; J. Pharm. Chim. *14*(3), 99-113; *14*(4), 154-161 (French).

BRIDEL, M., and LAVIEILLE, R. 1931B. The sweet principle of the leaves of Kaa-he-e (*Stevia rebaudiana* Bertoni). II. The products of enzymic hydrolysis of stevioside: Glucose and steviol. Compt. Rend. *193*, 72-74 (1931); Bull Soc. Chim. Biol. *13*, 636-655 (French).

BRIESKORN, C. H., and SAX, H. 1970. Synthesis of some derivatives of glycyrrhizic and glycyrrhetic acids. Arch. Pharm. (Weinheim) *303*, 905-912 (German).

DEUTSCH, E. W., and HANSCH, C. 1966. Dependence of relative sweetness on hydrophobic bonding. Nature (London) *211*, 75.

DORFMAN, R. I., and NES, W. R. 1960. Antiandrogenic activity of dihydroisosteviol. Endocrinology *67*, 282-285.

FINLEY, J. W., and FRIEDMAN, M. 1973. New sweetening agents: N'-formyl and N'-acetylkynurenine. J. Agr. Food Chem. *21*, 33-34.

FLETCHER, H. G., JR. 1955. The sweet herb of Paraguay. Chem. Dig. *14*, 7, 18 (July-August).

HOFMANN, A. 1972. A new sweetener of the indole series. Helv. Chim. Acta *55*, 2934-2940 (German).

HOROWITZ, R. M., and GENTILI, B. 1969. Taste and structure in phenolic glycosides. J. Agr. Food Chem. *17*, 696-700.

HOROWITZ, R. M., and GENTILI, B. 1971. Dihydrochalcone Sweeteners. *In* Sweetness and Sweeteners. G. G. Birch, L. F. Green, and C. B. Coulson (Editors). Applied Science Publishers Ltd., London, pp. 69-80.

INGLETT, G. E. 1971. Recent Sweetener Research. Botanicals. P.O. Box 3034, Peoria, Ill.

INGLETT, G. E., *et al.* 1969. Dihydrochalcone sweeteners—sensory and stability evaluation. J. Food Sci. *34*, 101-103.

JIZBA, J., DOLEJS, L., HEROUT, V., and SORM, F. 1971A. The structure of osladin—the sweet principle of *Polypodium vulgare* L. Tetrahedron Lett. No. *18*, 1329-1332.

JIZBA, J., *et al.* 1971B. Polypodosaponin, a new type of saponin from *Polypodium vulgare* L. Chem. Ber. *104*, 837-846.

JIZBA, J., and HEROUT, V. 1967. Isolation of constituents of common polypody rhizomes (*Polypodium vulgare* L.). Collect. Czech. Chem. Commun. *32*, 2867-2874 (English).

KOEPPEN, B. H. 1968. Synthesis of neohesperidose. Tetrahedron *24*, 4963-4966.

KRBECHEK, L., INGLETT, G., DOWLING, B., WAGNER, B., and RITER, R. 1968. Dihydrochalcones. Synthesis of potential sweetening agents. J. Agr. Food Chem. *16*, 108-112.

KURIHARA, K., KURIHARA, Y., and BEIDLER, L. M. 1969. Isolation and mechanism of taste modifiers: Taste-modifying protein and gymnemic acids. *In* Olfaction and Taste, Vol. III. C. Pfaffmann (Editor). Rockefeller Univ. Press, New York, pp. 450-469.

LYTHGOE, B., and TRIPPETT, S. 1950. The constitution of the disaccharide of glycyrrhinic acid. J. Chem. Soc. 1983-1990.

MARSH, C. A., and LEVVY, G. A. 1956. Glucuronide metabolism in plants. 3. Triterpene glucuronides. Biochem. J. 63, 9-14.

MAZUR, J. M., et al. 1970. Structure-taste relationships of aspartic acid amides. J. Med. Chem. 13, 1217-1221.

MOSETTIG, E., et al. 1963. The absolute configuration of steviol and isosteviol. J. Am. Chem. Soc. 85, 2305-2309.

NIEMAN, C. 1958. Stevioside. Zucker-u. Suesswaren-Wirtschaft 11, 124-126, 236-238 (German).

PILGRIM, F. J., and SCHUTZ, H. G. 1959. Cited by A. R. Lawrence and L. N. Ferguson. Exploratory physiochemical studies on the sense of taste. Nature (London) 183, 1469-1471.

POMARET, M., and LAVIEILLE, R. 1931. The sweet principle of Kaa-he-e. IV. Some physiological properties of stevioside. Bull. Soc. Chim. Biol. 13 1248-1252. (French)

RUZICKA, L., FURTER, M., and LEUENBERGER, H. 1937A. The empirical formula of glycyrrhetic acid. Helv. Chim. Acta 20, 312-325 (German).

RUZICKA, L., JEGER, O., and INGOLD W. 1943. New evidence for a different place of the carboxyl group in oleanolic acid. Helv. Chim. Acta 26, 2278-2282 (German)

RUZICKA, L., and LEUENBERGER, H. 1936. Glycyrrhetic acid. Helv. Chim. Acta 19, 1402-1406 (German).

RUZICKA, L., LEUENBERGER, H., and SCHELLENBERG, H. 1937B. Catalytic hydrogenation of the α,β-unsaturated keto group in glycyrrhetic acid and keto-α-amyrin. Helv. Chim. Acta 20, 1271-1282 (German).

RUZICKA, L., and MARXER, A. 1939. Conversion of glycyrrhetic acid into β-amyrin. Helv. Chim. Acta 22, 195-201 (German).

SCHLOESSER, E., and WULFF, G. 1969. Structural specificity of saponin hemolysis. I. Triterpene-saponins and -aglycones. Z. Naturforsch. B 24, 1284-1290 (German).

SHALLENBERGER, R. S., and ACREE, T. E. 1969. Molecular structure and sweet taste. J. Agr. Food Chem. 17, 701-703.

SHALLENBERGER, R. S., and ACREE, T. E. 1971. Chemical structure of compounds and their sweet and bitter taste. In Handbook of Sensory Physiology, Vol. IV, Chemical Senses; Part 2, Taste. L. M. Beidler (Editor). Springer Verlag, New York, pp. 221-277.

SINSHEIMER, J. E., RAO, G. S., and McILHENNY, H. M. 1970. Constituents from Gymnema sylvestre leaves. V. Isolation and preliminary characterization of the gymnemic acids. J. Pharm. Sci. 59, 622-628.

STOECKLIN, W. 1969A. Chemistry and physiological properties of gymnemic acid, the antisaccharine principle of the leaves of Gymnema sylvestre. J. Agr. Food Chem. 17, 704-708.

STOECKLIN, W. 1969B. Gymnemagenin, structure and O-isopropylidene derivatives. Helv. Chim. Acta 52, 365-370 (German).

TSCHIRCH, A., and CEDERBERG, H. 1907. Glycyrrhizin. Arch. der Pharm. 245, 97-111; Chem. Zentr. 1907(I), 1799-1800 (German).

TSCHIRCH, A., and GAUCHMANN, S. 1908. Further investigations on glycyrrhizic acid. Arch. der Pharm. 246, 545-558; Chem. Zentr. 1908(II), 1604-1605 (German).

UNTERHALT, B., and BOESCHEMEYER, L. 1971. Oximes as artificial sweeteners. IX. Unsaturated oximes. Z. Lebensm. Unters. Forsch. 147, 153-155 (German).

UNTERHALT, B., and BOESCHEMEYER, L. 1972. Cycloalkylsulfamic acids and their salts. Z. Lebensm. Unters. Forsch. 149, 227-229 (German).

VAN DER VIJVER, L. M., and UFFELIE, O. F. 1966. Presence of glycyrrhizin in rhizomes of Polypodium vulgare collected in Netherlands. Pharm. Weekbl. 101, 1137-1139; Chem. Abstr. 66, 52936 (English).

VAN NIEKERK, D. M., and KOEPPEN, B. H. 1972. Synthesis of 2-*O*-α-L-rhamnopyranosyl-D-galactose and some taste-eliciting flavonoid 2-*O*-α-L-rhamnopyranosyl-β-D-pyranosides. Experientia *28*, 123-124.

VIGNAIS, P. V., DUEE, E. D., VIGNAIS, P. M. and HUET, J. 1966. Effects of atractylligenin and its structural analogs on the translocation of adenine nucleotides in mitochrondria. Biochim. Biophys. Acta *118*, 465-483.

VIS, E., and FLETCHER, H. G., JR. 1956. Stevioside IV. Evidence that stevioside is a sophoroside. J. Am. Chem. Soc. *78*, 4709-4710.

VOSS, W., KLEIN, P., and SAUER, H. 1937A. Glycyrrhizin. Ber. *70B*, 122-132 (German).

VOSS, W., and PFIRSCHKE, J. 1937B. A new type of disaccharide as the sugar part of glycyrrhizin. Ber. *70B*, 132-137 (German).

VOSS, W., and BUTTER, G. 1937C. The isomerism of glycyrrhetic acid. Ber. *70B*, 1212-1218 (German).

VOVAN, L., and DUMAZERT, C. 1970A. The constitution of glycyrrhizic acid. I. Preparation of pure glycyrrhizic acid. Bull. Soc. Pharm. Marseille *19*, 41-44 (French).

VOVAN, L., and DUMAZERT, C. 1970B. The constitution of glycyrrhizic acid. II. Identification of the osidic part by chromatography. Bull. Soc. Pharm. Marseille *19*, 45-50 (French).

WOOD, H. B., ALLERTON, R., DIEHL, H. W., and FLETCHER, H. G., JR. 1955. Stevioside I. The structure of the glucose moieties. J. Org. Chem. *20*, 875-883.

YAMATO, M., *et al.* 1972. Syntheses of biologically active isocoumarins. Chemical structure and sweet taste of 3,4-dihydroisocoumarins. J. Pharm. Soc. Jap. (Yakagaku Zasshi) *92*, I. 367-370; II. 535-538; III. 850-853 (Japanese).

Index

ACKNOWLEDGMENTS

We would like to thank Susan Rabiner, our editor at Basic Books, for initiating and guiding this project, Pamela Hines for suggestions and advice throughout, and Sara Lippincott for turning an unruly manuscript into a trim narrative.